Reviews of the book *Digital Korea*

*"The two authors of "Digital Korea" draw attention to the dynamic changes taking place in Korea and introduce readers to digital lifestyles that will be experienced around the world in the near future. The 12 case-study reports are particularly useful to those who want to learn about the "Digital Nirvana" in Korea. "**Digital Korea**" is a must-read book showing us that the future of Digital Society is close at hand rather than an abstraction about a remote future."*

Seong-ju Lee, Editor-in-Chief **TelecomsKorea.com** South Korea

"For those of us working with the digital youth and struggling to understand the trends that are shaping our digital futures this book is priceless. A lot of what we in the UK think of as futurology is actually already happening in Korea. Packed with case studies and behavioural analysis this is a hugely rewarding read for anyone needing to gain insights into how digital society may evolve over here and how innovation is moving ever faster – 'bballi bballi'!"

Peter Miles, CEO, **SubTV** UK

"Having been involved with the planning and implementation of the Broadband and Telecoms revolution in Korea and internationally on Telecoms missions since, I am really impressed how Tomi and Jim have captured the essence and achievements of the Korea Digital journey from a range of cyber citizen, venture, corporate and Government perspectives. This book really helps the reader understand what makes a winning Digital ecosystem within a Global context"

JaeHong Yoon, Senior Vice President **Korea Telecom** South Korea

"I recommend this book as an insightful resource base for the near future concept creation, as the penetration rates of broadband Internet, 3G mobile and digital TV reach the levels in South Korea today."

Karri Mikkonen, Director of Strategy, **TeliaSonera Group** Sweden

*"Having worked with, admired and continually been amazed at the sophistication of the Korean telecoms sector, **Digital Korea** goes a long way to uncover and explain some of the secrets of this success and how it could influence digital futures internationally."*

Mark Newman, Chief Research Officer, **Informa Telecoms and Media** UK

"The book shows how intensely gamers become involved in virtual worlds and multiplayer gaming environments. The authors accurately describe how demanding the South Korean gaming environment has become due to the skilled professional gamers."

Alvin Yap, CEO **Nexgen Studio** Singapore

"Jim O'Reilly and Tomi Ahonen provide a great insight on Korea's hot bed of activity in Emerging technologies particularly Digital Convergence. It illustrates how Government policy combined with speed of economic growth , exports and fierce customer demands can shake up Service models and Industries globally."

Robert Jelski, Global Sector Head, **3i** UK

Copyright © futuretext Ltd 2007 　　　　　　　　　　Tomi T Ahonen & Jim O'Reilly

Digital Korea

Convergence of Broadband Internet, 3G Cell Phones, Multiplayer Gaming, Digital TV, Virtual Reality, Electronic Cash, Telematics, Robotics, E-Government and the Intelligent Home

By
Tomi T Ahonen and Jim O'Reilly

FUTURETEXT

Copyright © 2007 futuretext Limited
www.futuretext.com

Copyright ©2007 futuretext Limited
issue date 07.07.2007
Published by
futuretext
36 St George Street
Mayfair
London
W1S 2FW, UK
e-mail: info@futuretext.com
www.futuretext.com

All rights reserved. No part of this publication may be reproduced, in any form or by any means, electronic, mechanical, photocopying, recording or otherwise, without the prior permission of the authors.

Although great care has been taken to ensure the accuracy and completeness of the information contained in this excerpt document, neither futuretext Limited nor any of its authors, contributors, employees or advisors is able to accept any legal liability for any consequential loss or damage, however caused, arising as a result of any actions taken on the basis of the information contained in this excerpt document.

Certain statements in this excerpt document are forward-looking. Although futuretext believes that the expectations reflected in these forward-looking statements are reasonable, it can give no assurance that these expectations will prove to be correct. futuretext undertakes no obligation or liability due to any action arising from these statements. All third party brands and trademarks belong to their respective owners.

ISBN   978-0-9556069-0-8

Contents

Foreword by Dr Hyun-Oh Yoo, CEO SK Communications . . . . . page xi

Acknowledgements xiii

Chapter 1 - Introduction . . . . page 1
The Future Exists in South Korea
- World leader
- Seoul as digital nirvana
- Ubiquitous is the word
- We take survey of Digital Korea
- Like looking at still image of a movie
- How will it impact our lives?
- Society needs to change
- Attitude to technology
- A mirror into your own future

Chapter 2 - Digital Youth 15
Generation-C in South Korea

A - Lifeline is Text Messaging 17
- Generation Community is learned in class
- Friends that are always available
- Only SMS can do it
- SMS is preferred
- Adults don't get it
- Excuse to be rude
- Texting when bored
- SMS vs. email

B - Smart Mobs 22
- Mobile is the favorite toy

C - Me and My Peers 23
    Group Calls
    Will not speak on family wireline phone
    Taking the phone to bed with you
    Cameraphones
D - Smaller tribes 26
    Cellphone and identity
    Who may use the phone?
    Sharing, connecting with community
E - Natural Born Gamers 28
    Born to share
    Not like their parents
    Generate own content
    My phone my money
F - Simultaneous Parallel Networks 31
    Multiple connections, multiple communities
    Cool is digital cool
    The leading country, its pioneering users

Case Study 1 - Cyworld 34

Chapter 3 - Virtual Worlds 37
Our second lives

A - Growing Up with Virtual Reality 39
    Me and my avatar
    SayClub
    Mini me
    Virtual playgrounds
B - Cyworld 42
    Miniroom
    The Acorn, the business engine
    Music world
    What is your welcoming song?
C - Blogging 46
    Be my friend?
    Moblogging
    Cyworld in context
D - Friendships and Business 49
    30,000 businesses in Cyworld
    The emotional messenger - Nateon
    World invasion

E - Learning Through Virtuality 50
    Simulations
    Games and learning
    Fatigue and stress
    Growing up virtual

Case Study 2 - Lineage II 54

Chapter 4 - Intelligent Home 57
Digital living

A - Intelligent Home 61
    Security
    Remote control
    What of the pet
    And home wiring
    Intelligent door locks
    Is commercial reality today
B - Intelligent Clothing 63
    Dispersed computing
    Wearable computing
    Bridging the gap
    Digital monitoring
C - Near the Home 66
    Intelligent bus stop
    Ubiquitous dream hall
    Extends beyond the home
    To finish living

Case Study 3 - The Intelligent Mirror 71

Chapter 5 - Portable TV 73
Broadcast and beyond

A - A Leap Ahead 75
    Digital TV to cellphones
    Rich market for rich TV content
    Not your average pocket TV
    Not just snacking

B - The Six Legacy Mass Media   80
    First of the mass media: print
    Second media channel: recordings
    Third media channel: cinema
    Fourth media channel: radio
    Fifth of the mass media: TV the current giant
    Internet the sixth, is the first interactive media channel
C - Cellphone is Seventh Mass Media Channel   84
    Has abilities beyond first six mass media
    User-generated content
    Cannibalizing threat to older media
    All seven will continue
D - Is DMB part of Seventh Mass Media   86
    Interactivity is key
E - Other TV and Video Concepts   88
    Non broadcast video content to cellphones
    Mobisodes
    Simulcasts
    What of Video-On-Demand (VOD)
    Hana TV

Case Study 4 - Tu Media   92

Chapter 6 - Online Shopping   95
Money turning digital

A - Mobile Wallet   98
    Moneta
    No SIM card
    Do you want plastic with your card?
B - Other Electronic Payments   100
    Postal banking
    Cyworld is also a virtual shopping world
    Convergence with the finance industry
    How does it work?
    Automated digital interactivity
C - Online Auctions   103
    Skip the haggling
    Always digital
    Designed in ubiquity wins again

D - Communities Dominate 104
    No second hand
    Demanding customers
    Digital footprint
    Naver
    Silver digitals
    Family matters
    RFID in wrapping paper?

Case Study 5 - 2D Barcode 110

Chapter 7 - Electronic Government 113
Broadband bureaucrats

A - Starts with the Citizen. 115
    Digital society
    A cunning plan
    Increased private time
B - Educated Employees 118
    Teachers
    Empowered employees
    Labor
C - Innovative E-Government 120
    Taxes online
    Procurement
D - Legal Framework 121
    Data protection
    Coming back to the people
    Digital etiquette
E - E-Health 123
    Diabetes phones
    Smart floors
F - Local Government 124
    Collaboration is key
    SMS in politics
G - Law Enforcement 125
    Traffic wardens snap pictures
    Cyber security
    Viruses and malware
    Stolen phones
    Cyber terror

H - Where did the money come from? 129
   All part of the plan
   The future
   U School example
   What about the villagers?
   Conclusion

Case Study 6 Ohmy News 132

Chapter 8 - Machine Telematics 135
The intelligent automobile

A - The Next Internet is the Car? 137
   Car and web
   Where am I
   LG and GM
   Mozen
   TV in the car?
B - Portable devices 143
   Saving lives
   Telematics services and phones
   Mobile operators also in game
   Competition benefits
   How about pagers
C - Traffic and Parking 144
   Beyond paying for parking
D - RFID systems 145
   RFID applications
   How is your casino doing?
   Mobile RFID
   Future applications of mobile RFID

Case Study 8 - 3G Traffic Cam 150

Chapter 9 - Mobile Music 153
What after the iPod?

A - Ringing tones 155
   Not just the young
   Waiting tones (ringback tones)
   Soon bigger than ringing tones

B - MP3 Files 158
    Musicphones
    Apple's iPhone, not revolutionary in Korea
    Melon music
    Will not stop at sound of music
C - Make Your Own Music 161
    From music sound to music video
    Do it yourself music video
D - Enjoy Live Music 164
    Learn to dance
E - Cyworld and Music 165
    Background tunes
    Welcoming tunes
    Big ecosystem for digital music
F - Fighting Piracy 166
    Export of Korean music artists
    Current developments

Case Study 8 - Melon Music 169

Chapter 10 - Pervasive Technology 171
Ubiquitous Connectedness

A - Most Broadband 173
    Penetration
    Highest speeds
B - 3G Cellphones 176
    2D Barcodes
C - How Achieved? 177
    Geography helps
    Collective belief
    Virtuous cycle
D - Ideal Test Bed 179
    Western companies coming in
E - Privacy 181
    Selling sauna snaps
    Role of communities
    The end of eMail?
    Internet cafes
F - Cellphone the remote control 183
    Cellphone rules of conduct
    Lessons for the rest of the world

Case Study 9 - Wearable Computing 187

Chapter 11 - Multiplayer Gaming 189
Immersive entertainment

A - How big is big. 191
    Gaming generation
    Digital divide among gamers
    Impact of mobile gaming
    Immersive
B - Massively Multiplayer 194
    Massive realities
    Earn goods or purchase goods
    How big?
    Measured in real dollars
C - Enter the Colossus: Lineage 198
D - Casual Multiplayer 199
E - Professional Gamers 200
    World champions
    Farming for videogame gold
    Cheat at gaming logic
    Valued at 830 million dollars in Korea alone
    End of the game

Case Study 10 - Kart Rider 204

Chapter 12 - Consumer Robotics 207
One into every Korean home

A - Robot Means I Work 209
    Starts with the plan
    South Korean robotics
B - Household Robots 211
    Cleaning the home
    Teaching babysitter
    Networked humanoids
    Transformers come to life
C - Assistance Robots 214
    May I help you find your way?
    PGR the "female" assistant robot

D - Approaching Humans 215
    Ubiquitous Robotic Companion URC
    EveR-1 with humanoid skin
    Bringing humanity to technology
    Toy Robots

E - A Culture of Robotics 218
    Intelligent Robot Exhibition
    Robot Olympiad
    Future
    You will be assimilated

Case Study 11 - uPostMate 220

Chapter 13 - Digital Convergence 223
Internet, telecoms and media

A - Y of Convergence 225
    Three major elements converge
    Datacoms / the internet
    Telecoms / cellphones
    Broadcast / TV
    Center of convergence

B - Converged Solutions 232
C - Broadband Converged Network, BCN
    Fits the Y
    Not easy
    Technology vs business model
    Nespot Swing

D - WiBro 232
    Wireless and mobility
    Landline carriers
    Users appreciate mobility

E - Cellphones and Convergence 237
    Too fast
    Internet and cellphones
    Now all have it
    PBX replacement
    Counters VoIP and WiFi
    Convergence is so much more

Case Study 12 - Gigabit Broadband 245

Chapter 14 - Conclusion 247
Can Korea Maintain its Lead
      And access
      3G cellphones and 2D barcodes
      What of Japan?
      Its been our pleasure

APPENDIX: 253

Abbreviations

Bibliography

Websites

Blogsites

Index

About the authors

Other books by Tomi T Ahonen

Excerpt from book *Communities Dominate Brands* by Ahonen & Moore

Excerpt from book *Mobile Web 2.0* by Jaokar & Fish

Foreword

I recently finished an executive speaking tour in Europe and America, and have been struck by the increasing gap between the information society which has only started to emerge in Europe and America and how far the digital age has already become common place in South Korea. My peers are regularly surprised and curious to find out more about how Digital citizens are rapidly evolving and how living in the most connected culture in the world is quickly changing how we live and create.

Recognizing that this book will help educate citizens, leaders in business and government on what a real information society can be, I'm delighted and humbled to contribute to this book and be a fellow evangelist and champion of the power of digital citizens and technology sectors to make a difference.

We at SK Communications are extremely proud to play a part in contributing to the Digital Korea Blueprint with Cyworld, a service favorably used by more than 20 million Koreans. There are many ways of categorizing the service - from literally Cy Relationship world, to advanced social Media networking, next generation blogging, music distribution, Web 2.0, and not to mention video sharing. The fact is Cyworld, like many Digital Korea offspring is still evolving fast in a range of countries on different continents. Cyworld, as the most advanced community platform, is also converting millions of PC based users to mobile. It became the first mass market success story for monetization of social network services through closely integrating its service values with the right business models.

What really keeps us excited is the always-on-connection we have with the most digitally demanding audiences in the world and how that will shape the future of the Internet, Telecoms and even traditional media and Broadcasting Industries.

With the help of digital innovation and rapid development of the internet, the boundaries between industries and walls between digital devices are also collapsing. In this ubiquitous environment, services that can connect the users' life seamlessly, regardless of time, space, and device, will play the most crucial part. I can confidently say such services will be provided by internet platform providers. That is why the largest companies from various industries are struggling to take the lead in the internet platform field. Cyworld is the most advanced service in the world, where you can take a glance now at how the future will look when ubiquitous computing is fully realized. Moreover, Korea is 'the market' where such competition is most intense.

Like many of you reading this page I also still marvel and dream about how we best use Digital know how to further engage and make relationships with each other more rewarding and can proudly say the essence of Cyworld and Digital Korea will be a global and uninhibited inspiration for some time to come.

This book caters for both those already with some insight into Korea and also those learning and discovering Digital cyber athletes, secrets and relevance for the first time. The authors have done extremely well to position Korea against other World references in a balanced, simple and elegant way. Valuable to the reader is the thoughtful and comparative approach benchmarking the amazing Korea influences with other country achievements and allowing the opportunity to challenge and discover the relevance of the world class mass market case studies in an easy supportive way. This book is also a great independent reference for Digital vision and innovation globally.

Dr Hyun-Oh Yoo
President and CEO Korean SK Communications
Operator of online world, CYWORLD service

Acknowledgements

This book has been a wild, exhilarating ride across the width and the depth of the Digital Korea landscape. We have been fuelled equally by outstanding innovative venture and corporate successes inside and outside of Korea and the belief, vision and policies of the South Korean government and its advanced network and ecosystem of IT, telecom and broadcasting organizations. With the resulting introductions, meetings, and access to various specialists, research and journals, we have been exposed to the very bleeding edge of research and development in all areas of the digital spectrum. It is near impossible for us to mention everybody who has helped with this project. But we will try to acknowledge the most obvious people.

We want to start off by thanking both Hyun Jin Ko the previous president and Young Min Ryu, current president of KIPA the Korean IT Promotion Agency. We thank Sun Bae Kim president of the newly organised KIICA. We also received countless insights from programs committed to by the then Minister of Information and Communications the Honorable Dae-Jae Jin 2002-2006 and current Minister the Honorable Jun-Hyung Rho.

We thank Jason Lee, Peter Kim and Dal-Hyun Nam OnTimeTek, Woo-Jae Lee Infraware, Ki-Wan Park Iconlab, Yeon-Hak Kim and Paul Lee KTF, Peter McKinnon LG-Nortel, Harry Lee In Wireless, David Blundell NC Soft, Robert Lee and Won Kim Samsung, Hyeok-Jae Lee Mobile RFID Forum, Seong-Jae Lee InnoAce, Jeong-Sik Park TTA, Dr Anthony Park and Won-Chul Chang Samsung, Dr Chan Yeun LG Electronics, Sung-Ho Yoon KT, Brian Cho and Damian Kim SK Telecom, Hyeran In SK Communications, Darren Vogul and Sang-Hyo Kim Widerthan, Jason Hwang and Peter Kim Intromobile, Ho Song Neosol-tech, Reo Kim Darim and James Ahn INKA Entworks, Jae-Shob Shin Pixtree, MK Shin Anyfill, Leo Kang Tmaxsoft, Nina Han iRiver, Jay Sakong Goyang City, Thomas Cheung and Scott Gi KIPA, Matthew Weigand Korea IT Times, Jun-Hee Shin and Seung-Woo Choi KIICA, Mervyn Levin DTI, Chang-Eun Park, Paul and Carol Deeley, Dozie and Chi Chi Azubike.

We are grateful to Young-Sun Soh British Embassy Seoul, Mark Newman Informa, Matthew Weigard Electronic Times, Scott Best Lehman Brothers, John Slamecka AT&T, Jennifer Schenker Red Herring, Gary and Paula Wassell, Jean Pierre and Eun-Wha Van Ovost, Graham and Doreen GrahamKerry Associates, Andy Eccles, Martin and Theresa Kelly, Jon Richard, and Paul Robin Birmingham Crew Kevin & Dwyers, Bakers, Considines, Kellys, Copes, Dheers, McDonalds and O'Reillys worldwide. Thanks to Kim-Young Shin YongIn City Digital Projects for his global insight, intellect, and determination to grow the Korean Digital economy.

We acknowledge Peter Miles SubTV, Steven Chan AMI, Voytek Siewierski Mitsui, Lars Cosh-Ishii and Daniel Scuka Wireless Watch Japan, Tony Fish AMFVentures, Howard Rheingold Smart Mobs, Alan Moore SMLXL, Mark Curtis Flirtomatic, Neil Montefiore and Claudine Lim M1, Walter Adamson Digital Investor, Benjamin Joffe Plus8Star, Rory Sutherland Ogilvy, Nic Frengle IntaDev, Mitsuro Tada Daito, Toshikazu Tanida CIAJ, Shigeki Ishizuka NTT DoCoMo, Olof Schybergson and Mike Beeston Fjord, Jouko Ahvenainen Xtract, Leong Chou Weng Singapore Infocomm; Kenneth Chang MiTV, Alfie Dennen MoBlog UK, William Volk MyNuMo, Jackie Danicki Engagement Alliance, Heikki Karimo IBM, Alex Tan, Russell Anderson, Jochen Metzner, Nicole Cham, Janne Laiho and Ebba Dåhli Nokia, Olav Henrik Kjorstad Telenor, Judi Romanchuk, Kari Onniselka Talent Partners, Rick Pryor and Stefan Ciesielski Siemens; Blums Pineda Globe, Russell Buckley and Carlo Longino MobHappy, Mike Short MDA, Roberto Saracco TIM, Steve Flaherty, Claus Nehmzov Shazam, Adriana Cronin-Lucas Big Blog Company, Peter Holland and Anthony Santiago Oxford University and all of ForumOxford. And we cannot forget Paul May, Paul Golding, Sara Melkko and Minna Rotko.

As our special guides to the minds of Generation-C we want to thank Jon and Ere Luokkanen, Olli and Salla Kasper, Ema and Joseph Moore, and the prototypical Gen-C: Taina Kalliokoski. We also thank our friend Ajit Jaokar of Futuretext for his patience and help in the planning and production of this book.

Finally we want to thank our families for their continuing support in this demanding project, as this book was written outside our hectic daytime schedules in between many international events, projects and family commitments. Sincere thanks to Kay and Flan O'Reilly for making Jim and fellow brothers and sisters Tom, Lynn, John, Sue, Catherine, Nick, Alice, Neil proud and prepared to challenge the boundaries of life. The deep love and support of Kay for our father Flan every day without complaint. Heartfelt thanks to Alice and Neil Baker and the late Kieran O'Reilly for their limitless family support and Flan for his dignity, perseverance and inspiration for a more humane and technology-assisted world for all.

To Jim's darling wife, heart and soul Doctor, June-Young O'Reilly for her encouragement and sacrifices to help finish this project and yes we will still dance at dawn every day when we are eighty years old.

And as this is a book about our digital future Tomi wishes his best to the next generation Maria and Katariina Karimo; Luca, Leo, Timotei Lundgren; Iiris and Aamos Lundgren; Roni, Kris and Petteri Jorgensen.as well as Jim sending best wishes to Conor, Niamh, Caitlin, William, Zimie, Charlotte, Alex, Oliver, James, Grace, In-Ha, So-Ha, Woo-Ju, Ae-Ri, Jon-Su and Woo-Ram

Jim would like any resulting gains from this book to contribute to Professor Kil Sung Oh and Do Young Oh for their outstanding efforts for the Korean Disabled at the Assistive Technology Research and Assistance Center (www.atrac.or.kr) and Il Ha Yi , President at to the Korea based International Humanitarian Organisation Goodneighbours International (www.gnint.org).

Materials relating to this book will be added to website **www.digitalkorea.info**
We welcome feedback for this book, please send to **feedback@digitalkorea.info**

Copyright © futuretext Ltd 2007   Tomi T Ahonen & Jim O'Reilly

Chapter I
Introduction

The Future Exists in South Korea

Image courtesy *IT Korea Journal*

2

> *"The empires of the future are the empires of the mind."*
> **Sir Winston Churchill**

1
Introduction
The future exists in South Korea

What is the future like? We know that current trends suggest the internet will be available practically everywhere. We know that internet access is becoming increasingly broadband. We know broadband speeds are climbing. In addition, we know internet users are increasingly accessing the web wirelessly. We know cellphones reach ever more of the population. We know cellphones are going to high speeds of 3G and beyond. We know that media is going digital, from music to movies and digital TV. We see convergence of media, web and communications, from such innovations as IPTV to services such as Skype and Vonage. But where will it all lead?

What will the world be like in the near future, when everybody has internet access, at high-speed broadband? When everybody has not only a cellphone, but also even teenagers have cameraphones on high-speed 3G networks? How will our daily life be different when our home is digital and connected, when all government services are provided digitally and when digital TV is in our car, on our laptop and on our phone? When virtual worlds are formed, digital communities emerge, our money is digital, and our online lives take on a meaning of their own? What happens to our work, our families, our lives?

This is a book about that future. It is not science fiction. That future is already now visible in only one country, South Korea. The country with the highest penetrations of wireless broadband internet, 3G mobile cellphones, portable digital TV, online virtual gaming and so forth. To see the future, you need to understand South Korea, or like we say, Digital Korea. Let us take you on a tour.

World leader

90% of South Korean homes have broadband internet access. The world average is about 20%. 63% of South Koreans make payments using their cellphones, the world average is under 5%. 43% of South Koreans maintain a blogsite or personal profile online like Americans might have on MySpace or the British on Bebo - the industrialized world average is about 10%. Over half of South Koreans have migrated their cellphone account to 3G (the world average is 5%).

100% of South Korean internet access has migrated to broadband (the world, about 30%). 25% of South Koreans have played the same multiplayer videogame (*Kart Rider*, a case study later in this book). 40% of South Koreans already have a digital representation of themselves, a so-called avatar such as Western people who use Second Life or Habbo Hotel. By every measure, South Korea leads the world in digital adoption and innovation. In a very literal sense, South Koreans are living in the near future of the digitally converging technologies from the viewpoint of the rest of the world.

This is a book to show to Western readers how incredible that future is, what is now commonplace in this melting pot of digital innovation. Moreover, to show glimpses of where South Koreans see their future going. In ten years, every South Korean household will have a robot. Yes, in ten years. Does that make you pause? Read on...

Seoul as digital nirvana

For most South Koreans today, their very lifestyle already revolves around services that require high-speed digital access. The networks are state-of-the-art wireless networks and handsets: both international standards of 3G networks have been launched (CDMA 2000 1x EVDO like in America and WCDMA or UMTS like in Europe). Faster 3.5G networks on HSDPA are already in commercial use. WiFi and WiBro (a variant of what is called WiMax in the West) is commercially available, as is DMB digital TV broadcasts to cellphones, cars, laptop PCs and other movable devices.

WiFi is offered on the world's most extensive broadband wireless network and can be accessed from 25,000 cyber cafés located around the country. On broadband not only have all South Korean internet access lines been migrated from narrow-band (dial-up) to broadband, South Korea leads the world in broadband adoption as well as broadband speeds offered. Where most industrialized countries offered broadband speeds in the 2 Mbit/s to 10 Mbit/s range at the end of 2006, in South Korea the standard offering is between 50 Mbit/s and 100 Mbit/s and South Korea is already rolling out the first pilot connections of Gigabit broadband (1,000 Mbit/s). And to add insult to injury, South Koreans pay for their broadband at the lowest rates in the world.

In a very literal sense, to be South Korean means to be connected.

> **To be South Korean means to be connected.**

South Korea has usually been it the shadow of the economic "miracles" of its bigger Japanese neighbor, from home electronics, the camera and wristwatch industries, to automobiles and robotics. But with the dawn of the "Connected Age" the new wirelessly connected society which has moved beyond the "Networked Age" of the 1990s, the South Korean government, industry and academia all decided to pool resources and make a leap ahead of its Eastern neighbor across the sea.

A key ingredient is the Korean phrase "bballi bballi" which means to "hurry hurry". South Koreans work very hard for long hours, and this work ethic, together with the pooling of all resources have enabled a massive jump in the digital infrastructures. These in turn have helped propel the whole country and all other forms of society and economic life to gain from the lead in digitalization.

Ubiquitous is the word

The word you hear currently from all South Korean executives, officials and experts in the digitally converging industries is "ubiquitous". Ubiquitous computing, ubiquitous internet, ubiquitous gaming, ubiquitous coverage etc. Ubiquitous means "ever-present" or omnipresent; that something is everywhere, accessible at any location. Oxygen is ubiquitous on our planet but water is not (consider the Sahara desert for example). South Koreans are now on the verge of the world's first society where digital services are literally ubiquitous.

Broadband Penetration Per Capita

Fifteen countries with highest broadband penetration as percentage of population in 2006, according to the ITU

Country	%
South Korea	51%
Hong Kong	33%
Japan	33%
Italy	29%
Sweden	28%
Netherlands	27%
Denmark	27%
Iceland	26%
Switzerland	25%
Norway	24%
Finland	24%
UK	23%
Austria	22%
Canada	21%
Taiwan	21%

Source: ITU Digital Life 2006

We take a survey of Digital Korea

Jim lives in the UK and Tomi lives in Hong Kong. Both of us have regular business contacts with South Korea and have visited South Korea countless times. We research and discuss South Korean digital innovations regularly in our daily work both of us within the converging IT/telecoms/media industries. This book was intended to be a guidebook and overview of how advanced the society of Digital Korea is overall. What kind of synergies and serendipities start to happen when several closely related industries achieve the global cutting edge - or indeed the bleeding edge - of technological advances.

When broadband internet leadership is merged with multiplayer videogaming leadership and 3G mobile telecoms leadership and digital TV leadership, then tremendous advances from one will influence the other and so forth.

We start by studying the South Korean consumer, and examine how far Generation-C (the Community Generation) already exists in South Korea. We discuss several phenomena around youth and ubiquitous digital services. From cheating in school by using cellphones to spending hours inside multiplayer games to sending 100 text messages per day, we dig deep into the life of what being young, connected and Korean means today.

> **Pearl - Use your cellphone to learn.** Need to learn a new language, or how to program computers or help with your mathematics in school? South Koreans use hundreds of self-training services on their cellphones from learning the latest dance moves to web design.

Then we start to examine various service areas. We look at the role of virtual realities such as recently very much in the press in the Western World as Second Life has reached 2 million users. Contrast that with virtual worlds in South Korea such as Cyworld and Kart Rider both with over 20 million users exploring virtual world entertainment and life; even employment. We then examine digital living and the intelligent home. For all the success of Western massively multiplayer online games and environments from Everquest to CounterStrike and World of Warcraft, with its 7 million gamers worldwide, the South Korean game Lineage is the world leader with over 14 million registered gamers inside the massively multiplayer gaming environment.

We discuss the rapid adoption of digital TV tuners into cellphones and cars in the portable TV chapter, and then look at how shopping and banking are evolving electronic cash, credit cards embedded to cellphones and digital connectivity is omnipresent. We discuss for example what many call the biggest change in consumer behavior, 2D Barcodes - also first introduced in South Korea and now becoming such a big hit in Japan and global giants like Google and Nokia betting on this as the next big breakthrough. We offer 2D Barcodes as one of our twelve case studies for you in this book.

We look at healthcare, schools, law enforcement and such matters in the government chapter - for example that the opportunities for cybercrime

---

2006 in the USA 10% of music sales was digital
Source: IFPI January 2007

---

have caused every South Korean police department to set up a unit for cybercrime. We look at telematics, in particular as it relates to automobiles.

In addition, we survey the music industry in Digital Korea, to see why 45% of all music sold in South Korea is already delivered to cellphones or so-called musicphones, which were invented in South Korea with the first commercial launch in 2003. For contrast, the Apple iPod has existed twice as long as, yet even in America, at the birthplace of the iPod by end of 2006 only 10% of American music was sold online. The iPod was indeed revolutionary, but it had to thrive in isolation. The musicphones in South Korea benefited from the whole society being digital and connected, including the record labels and the very musicians themselves.

We continue by examining pervasive computing, multiplayer gaming, and finally perhaps the most amazing of them all: the South Korean intention to make consumer robots equally ubiquitous. Where most other countries cautiously consider bringing robots in touch with humans, in South

Korea the government is certain it will have a robot in every Korean home in ten years! So the robotics industry is rapidly developing the human interactivity knowhow in what guiding might a robot help with at a shopping mall, or how a household robot could also double as the robotic nanny to help the children finish their homework. We end the book with a discussion about convergence from technical, industry and device points-of-view before giving our summary and concluding thoughts.

Like looking at a still image of a movie

In some ways, our task to describe the continuously connected Digital Korea is nearly impossible. It is like attempting to describe a great movie by showing still images of the movie. Yes, these are the actors in these kinds of scenes, but the real impact comes out of the whole coming together, from the script to the actors to the scenery to the music, lighting, etc. We have strived

---

2006 in South Korea 57% of music sales was digital
Source: IFPI January 2007

---

to cover all of the most relevant points so that at least a fair view can be made. South Korea is not "only" the most connected country in broadband internet user, or in 3G cellphones, etc. We hope to give you a context.

For our book, we had the privilege of visiting with numerous South Korean pioneering companies and interviewing dozens of leaders in this space. Some of those discussions have turned into case studies but others have their thoughts woven into the text and provide more of the detail in the book.

Inevitably, with a society changing as fast as the bleeding edge of technology, many of the statistics in this book will become obsolete soon after we have gone to print. With that, we have attempted to give context of the statistics, comparing South Korea to the world average for example, or to use leading countries like Japan, USA, Canada, UK etc to provide comparisons and contrasts.

We hope these help to illustrate that it is not only one or two areas where South Korea has "sneaked" ahead; it is in just about every possible measure of a digital country or information society.

For you as the reader, please do not contrast our stated statistic with the latest data you find in your country when you read this book. Consider rather, how far ahead South Korea was in 2006 at the time of our research and writing of this book, and please take it as a given, that South Korean industry leadership will have moved on further since we wrote this book.

How will it impact our lives?

What will happen in this kind of environment, is a "virtuous cycle" of technical leadership. As the technical infrastructure is ahead of the world, it enables newest handsets and computers to be deployed, which in turn enables programmers to exploit to the fullest the abilities of the technologies, which in turn give the consumers choices that are innovative. This then results in customers who are familiar with the new, and thus become very sophisticated and demanding of ever more. They are willing to pay for cutting edge solutions, which again fuels further investments in faster technologies, etc. This is why so many usage numbers seem mind-boggling to Westerners.

For example, picture messaging from camera phones. In most Western countries the usage of "MMS" (Multimedia Messaging) is a severe disappointment at most networks. In the UK, with about 72 million mobile phones of which about 60% are cameraphones and 15% are 3G high speed phones, picture messaging traffic is under a million picture messages per month. In South Korea, with high resolution cameraphones, near 100% cameraphone penetration, high speed 3G networks, and services serving the picture sharing needs of its customers, like Cyworld, out of a population of 40 million mobile phone owners, they send 6 million picture messages per month. A user population of nearly half, sends six times as many.

Why is this? Part of it is that everybody has the devices, and they all have high-speed access. That is only part of the story. The bigger part is the benefit from new, related innovations, only possible in that environment. A staggering statistic is that 90% of the picture messaging transfers are in fact cameraphone pictures uploaded to Cyworld. (For a Western context, imagine uploading directly from cameraphones to picture sharing online sites like Flickr or videos to YouTube).

Cyworld in South Korea is the world's most advanced digital online environment, which we discuss in several chapters and its own case study in this book. The virtuous cycle of devices, users, high-speed networks and most importantly the new services such as Cyworld, cause both a "push" and a "pull" effect, dramatically accelerating the usage, and improving service development.

Society needs to change

While technology allows new ways to do things, it invariably brings new problems with it. The same cellphone that allows us to contact our spouse urgently on the road, is also causing many drivers to shift focus from the road and causing accidents. The efficiency of rapid broadband access to social networking websites is good for publishing blogs and journals and joining in discussion groups. However, that same technology allows unscrupulous predators and sex offenders to target young innocent users and make untoward advances.

Issues of privacy arise, as do those of content ownership. The girl dancing in a video shot on her friend's cameraphone and uploaded on a video sharing site, may be original creative expressive art, but if she is dancing to a tune by Madonna - without permission - then it is likely an infringement on Madonna's rights to the song. New technologies bring with them new opportunities but also new threats.

South Korea has been at this frontier as well. Its society has been learning and adjusting, to issues ranging from students cheating in class by sending text messages during exams, to teenagers sending or receiving pornographic images on cameraphones, to privacy issues around tracking the locations of people by their phones, etc. Typical problems include the use of cameraphones in public baths, swimming pools, gyms etc. Some of the solutions around these kinds of problems can be technical - such as creating a sound to a cameraphone that it cannot be used to secretly take photos - to regulatory, such as forbidding the use of cameraphones at public baths and swimming pools - to behavioral, such as the guidelines for considerate phone usage, that have been published in South Korea. We discuss these issues as well in this book.

Attitude to technology

South Koreans today can be said to be immersed in information. The amount of information available to South Koreans at any time of any day from anywhere can be overwhelming. Smart phones, internet cafes, gaming halls, and restaurants with broadband access all entice South Koreans to keep access information and refresh their contacts. This vast access to data seems to fit naturally with the "bballi bballi" culture of hurry-hurry we discussed earlier.

Yes, everything is a rush. South Koreans want to be ever more productive. And, to save time and money in the process. Data and telecoms networks both those in use, and their higher speed siblings being deployed, have indeed expanded productivity. However, bballi bballi also has some drawbacks. In Digital Korea, people often will not take time to relax and to disconnect. In the rapid pace life in South Korea, even leisure activities, such as gaming, happen at a high speeds.

As the environment changes, we also change with it. The consumer in South Korea has been pampered with the world's widest range of digital services and choices. With it comes a maturity to such options and services. In our interviews for this book, Harry Lee, Vice President at In Wireless of South Korea, said:

> "The South Korean customer is not satisfied with voice communications only anymore. The customer wants life to be made easier. They are not satisfied with basic services such as text based data communications either, now they demand multimedia functions."
> Harry Lee, Vice President, In Wireless

We can see how this increasing awareness of what is possible results in ever more demanding customers, who then push the providers to supply ever better services.

A mirror into your own future

We hope you will enjoy this journey with us to Digital Korea. We are convinced that these changes will happen in all industrialized countries. With that, this book can serve as your digital guide into the next five years or so of our life, your career and employment opportunities, your family and home, etc.

If you work in the digitally converging industries like telecoms, internet, media, banking etc, we strongly recommend visiting South Korea to see it for yourself. Even to us, who discuss South Korean inventions and industries in our daily technology expert lives, we are still continuously surprised and amazed at the latest innovations and developments. The synchronized simultaneous lead in all of the digital industries is a virtuous cycle, which currently perpetuates an ongoing leadership in all areas of digital life.

To understand our digital future, understand Digital Korea.

Chapter II
Digital Youth

Generation-C in South Korea

Image courtesy *IT Korea Journal*

> *"The youth of today? With their constant text messaging contact, they **are** the Borg"*
> **Peter Miles, CEO SubTV**

II
Digital Youth
Generation-C in South Korea

Generation-C as the Community Generation was introduced in the book *Communities Dominate Brands* (Ahonen & Moore, 2005). The defining characteristic of Gen-C is that for the first time in mankind's history a new generation is growing up with permanent, 24-hour support of the friends, colleagues, community. The umbilical cord or the "lifeline" for Gen-C is the cellphone and the secretive cryptic connection method is SMS text messaging. It is not the only way Gen-C connect, in fact, the youth of today is inherently multitasking on multiple platforms from videogames to social networking sites to blogging and instant messaging. Nevertheless, one form of communication reigns supreme for Gen-C, and that is SMS text messaging.

A LIFELINE IS TEXT MESSAGING

Nowhere on the planet is the appeal of SMS text messaging more visible than in South Korea. According to the *Korea Times* of 9 February 2006, already a third of South Korean students send 100 text messages every day. For comparison, the world average is just over one SMS per day and in most markets, a level of ten SMS text messages sent per day is considered heavy usage.

Generation Community is learned in class

The ironic twist of the learned habits of Generation-C is that it is learned in school. Not that SMS text messaging is "taught" in school. Gen-C learns to send text messages very rapidly, secretively, and without even looking at the phone - from attempting to cheat in class. This is not by any means a Korean or Asian tendency; it is universal. Teenagers all around the world from Scandinavia to Canada admit to having attempted to send text messages during class. A survey by the Korea Agency for Digital Opportunity and Promotion of 1,100 Korean youth aged 14-19 in November 2005 revealed that 40% of Korean youth admit to sending text messages in class.

Friends that are always available

Generation-C is the first generation to live with the friends "in their pocket" - instantly available at all times. Again, it is important to highlight how different this is from any other communications. It is not the same if the

> ## European youth replace cellphones every 21 months
> Source Telephia Q1 2006

friends are there to support you on an online community like an instant messaging system, or in a chat room. Nor is it the same as friends who are available via e-mail or who can be called at home via a fixed wireline telephone. That friend may want to support you always, but the connection is not constant.

The reality of any other form of contact, is that it is not "always". The change in behavior that arises with Gen-C is that at any time in life, whether you are in the middle of an argument with your parents at the dinner table, or in your own bedroom crying because you broke up with your girlfriend/boyfriend, you can contact your friends via SMS text messaging: and they will be there.

Only SMS can do it

Only via SMS, text messaging can friends give instant support at any time. You cannot make a phone call in such circumstances, such as sitting at the dinner table with your parents or in a meeting with your boss. Nevertheless, you can send SMS text messages in those situations. A UK survey by YouGov and Carphone Warehouse in September 2006 revealed that 48% of British teenagers regularly send text messages to another person while talking to someone else.

This is absolutely vital in understanding the "umbilical cord" nature of SMS and how different Generation-Community is from you and me. An urgent message from a friend has to be responded to, nearly immediately. Else, the whole friendship is in trouble. Research in Japan's Keio University by Dr Fujimoto and Dr Ito, revealed that when young people send an urgent text message which is a call for help from a close friend, it has to be replied to within 30 minutes. In addition, being in a meeting with your boss, taking an exam in school, or asleep for example, is no excuse. Similar findings from

---

South Korean youth replace phones every 11 months
Source NIDA Sept 2005

---

the Catholic University of Leuwen in Belgium showed that more than half of teenagers have been awoken by a friend sending a text message at night.

SMS is preferred

SMS is the preferred means to connect for the youth, and "of course" it is ok to send a message to your friend while talking to someone else, like your parents, your teachers or your work colleagues.

This is radical change in behavior. Generation-Community will consult with the mates on the cellphone **before** and **during** consumer decisions, such as deciding which bar, club, disco or pub to go to tonight, or while in an electronics store, a member of Gen-C will consult with the electronics "guru" friend about some new product that is on sale. They will think nothing of haggling with five car dealers about the final price of a car - via text messaging. This generation is accustomed to living experiences together with mates, not in isolation. If the woman in the above example were part of Gen-C, she would have told her "shopping friends" about the trip to the mall before going, and sent some messages, or made calls, while in the store. At the coffee break, again she would advise her friends what was going on, and ask for advice on what was next on her mind.

Youth Cellphone Experiences 2006

Survey of 1,258 teenagers in Britain aged 11-17 by YouGov and Carphone Warehouse in 2006 found that teens reported the following experiences:

Will not let parents snoop inside phone	68%
Are allowed to stay out later because of carrying a cellphone	53%
Send text messages while talking to someone else	48%
Avoid contact by parents to their phone	37%
Communicate via phone with someone their parents would not approve	35%
Have had their phone stolen	11%

Source: Carphone Warehouse Mobile Life Youth Report September 2006

Even more, it is not only *you* who has to have access. For Gen-C to emerge, your friends also have to be permanently connected. Imagine if everyone sat at a laptop and wireless connectivity every day everywhere. They could evolve to Gen-C. The same tends to happen with teams where all have Blackberry wireless email devices. However, if the other friends are not permanently connected, they will not be there to provide the support under all conditions. Only the cellphone or of course other cellular network "always on" allow this.

Adults don't get it

The heart of a young person's phone is SMS text messaging. It is the mischievous and frivolous use of messaging used in this way. There is a counterculture for the young to have cryptic messages that adults are not even supposed to understand. Texting supports the need to be creative, to have entertainment - jokes - during the day.

> **40% of South Korean youth send SMS text messages in class**
> Source: Korea Agency for Digital Opportunity & Promotion 2005

     Where adults may think 160 characters to be terribly limiting, teenagers will find it funny to fit humorous, private and cryptic communication into the message.
     (Incidentally, the above paragraph is exactly 160 characters in length, the limit of a standard SMS text message). SMS is ultimate youth communication, fast, no grammar, limited characters and secretive. Yet it is used to create love notes, poetry and jokes. It is a natural evolution of language to fit a new technology. In New Zealand, some middle schools have already allowed students to use SMS grammar in their school homework and essays.

Excuse to be rude

Or to put it another way, the advertising giant OgilvyOne's global creative director, Rory Sutherland says, *"SMS text messaging is an excuse to be rude in communications."* With older formats of written communications, the letter, the fax and the email, we had to compose the communication. Start with a greeting, end it with a closing. Not in SMS. With text messaging there only is 160 characters, so we skip the politeness, and get straight to the point. An excuse to be rude. *"Can u meet at 8?"*. Very abbreviated communication for the generation that has the shortest attention span ever.

What may seem to many of us as the particularly strange behavior of our own children, is in fact a universal trend all around the world. The young generations show a clear preference of SMS as their communication channel of choice. Several studies around the world show consistently that the young users prefer SMS to voice calls and prefer SMS to e-mail. It is their private and secret communication tool that they can privately tap via their personal keyboards, without parents or siblings getting to listen in to a voice call or view an e-mail. Even America is following the trend, by the Autumn of 2006, the percentage of American cellphone users sending text messages had reached 42% according to the CTIA.

Texting when bored

Returning to South Korean youth and SMS. It is definitely the preferred means of communication. 40% of South Koreans send text messages when they are bored according to the Korea Agency for Digital Opportunity & Promotion Survey of 2005, covering 1,100 South Korean youth between ages 11 and 19. The same survey found that 20% take their cellphones to the bathroom and on the downside, 20% have received threatening messages via SMS.

SMS vs. email

Most of all, where they have access to both, young users in study after study after study confirm the same conclusion - even though e-mail is free and SMS text messaging costs, **all young users prefer text messaging** on SMS over e-mail. Korean young adults put it so well - email is outdated, it is not used between friends, and not between colleagues. The only people you would use mobile email with are the older generation at work, i.e. your boss. email? Its so 1990s.

Yet SMS text messaging is not the only way to connect. Generation Community is also very adept at using IM Instant Messaging and various chat rooms and the chat function of online social networking sites etc. We will examine online social networking in the Virtual Worlds Chapter next in this book.

B SMART MOBS

Beyond just text messaging, cellphones are transforming our behavior and introducing an unplanned herd behavior element. Howard Rheingold, the pioneer of virtual communities never loses sight of the primary purpose of the cellphone. Other uses are coming such as shopping and consuming content, but the power of the cellphone is communication. In enabling cooperation, as he says of the cellphones in his book *Smart Mobs*:

> *They amplify human talents for cooperation. They also change the way people shop, how they gather information on products they want to buy and where they decide to make that purchase.*
> Howard Rheingold, *Smart Mobs*, 2002

The Community Generation can be said to be very superficial, and to neglect "serious" and credible information sources on behalf of the first hits on Google and the most prevalent opinions found on the web. They also will steer towards the free sites rather than anything requiring subscription or payment, and this may limit access to some information sources or even color their views. These are longer-term trends that we cannot yet fully determine, but time will tell.

Mobile is the favorite toy

The cellphone is the most critical toy and tool for youth of today. Yes, they may love their iPods and Playstation Portables, and spend countless hours on their laptop and on the web in chat rooms and using IM, but if forced to select only one, increasingly the tool of choice is the cellphone. Again, this is a universal trend.

Even late at night, when parents tell their kids no more TV, no more internet, no more Playstation and turn off the music, the lights are turned off in the children's' bedrooms but the communications continues. The kids take their cellphones to bed and continue, silently into the night. Their favorite late night "fix" is that of text messages on the cellphone.

C ME AND MY PEERS

The cellphone has become the must-have device. It is seen as a major element in the definition of a young person's emerging persona. My phone. What it tells others of me. The need to have not only a cellphone, but also the right type of phone of the right brand, is following time-honored patterns seen with earlier generations of having the right brands of jeans or sneakers. Some sociologists in the UK have shown that what earlier generations associated as factors of being cool in smoking cigarettes is now seen through the cellphone. The Mobile Youth study of 2005 revealed that young people spend less on cigarettes, alcohol and chocolates because their money goes into their phone bill.

In South Korea, teenagers will replace phones every eleven according to the latest figures by NIDA, the global average replacement cycle is almost twice that at 18 months. That helps show how strongly South Korean youth value their cellphones. As increasingly kids all around the world have phones as their toys, they are now used in playing. Then as kids grow up, they become a part of flirting, dating etc.

Group calls

One of the strange types of "playing" with cellphones is the group call. A group of 4 teenager boys might call a group of 4 girls. Partly they are just friends, but there can easily be some bubbling romantic energy between one of the boys and one of the girls. They aren't dating yet, and this is a kind of "pre-courting" stage, when the interested boy and/or girl can be charming

---

10% of UK cellphone owners download games
Source: m:Metrics 2006

---

and funny and witty, while within the support of the close mates. He/she does not have to take the full responsibility of the phone call and get too deeply into a "relationship" but can still flirt with the love-interest. This all is of course via the speakerphone feature on the phone.

As a side comment, we do want to point out that the cellphone manufacturing industry never assumed that the speakerphone feature would be anything other than a business application on the high-end smartphones. How wrong we were. The biggest use of the speakerphone feature is by kids and teenagers playing.

Will not speak on family wireline phone

Another change in the behavior of the current generation and its predecessors, is the use of a fixed landline at home. As soon as teenagers got their own cellphones in all countries, they follow the same pattern. They won't answer the family fixed wireline phone - after all, any close friends will have the cellphone number, so a call to the fixed landline phone will "not be for me" but rather the teenager feels they have to behave like the family answering service, to take a call for someone else. Moreover, they don't place calls on that phone. The youth behavior trait was first chronicled by Keio University of Japan professor Mizuko Ito who says that kids use their personal phones at considerable cost, rather than the family landline phone, which tends to be free, just to avoid being overheard by their parents.

Taking the phone to bed with you

The global survey by BDDO in April 2005 revealed that 60% of cellphone users worldwide take the cellphone to bed with them. Physically to bed! In addition, the Spring 2006 survey by Nokia found that 72% of us use the phone as our alarm clock, meaning therefore that 12% of us - the older population no doubt - place our phones on the bedside table next to our bed. But nearly three out of four people sleep with the phone within arm's reach

---

37% of South Koreans download cellphone games
Source: NIDA 2005

---

that is how intense our relationship has become with the cellphone.

Practically all teenagers take their phones to bed of course. They will not be calling their mates, but for the next hour or two, they still send SMS text messages with the phone on silent. The first university study of SMS and cellphone addiction at the Catholic University of Leuwen in Belgium found in 2004 that more than half of all teenagers have woken up to an SMS sent by a friend at night - and 20% regularly do so. In fact, kids wake up more readily to the sound of an incoming SMS text message, than the sound of the wake up alarm on their cellphone.

Cameraphones

We adults like to think of the camera on our cellphone as being a variation of the older film-based cameras we grew up with. Back then film was expensive, prints were expensive, it took days to get film developed into pictures, and cameras themselves were expensive, as was the wide range of accessories. Pictures were not "wasted". We took the camera to important events like weddings and holiday trips etc, and any pictures we took we would then store and expect to see years later as mementos.

> **Pearl - use camera to "type" web addresses.** 2D Barcodes in South Korea are now used by majority of cellphone users to access websites. Rather than typing, the cameraphone owner just points the camera at the 2D barcode and the phone detects the text embedded in the barcode.

Not teenagers today. They never knew of film-based cameras. To them cameras were always digital. In 2001 Japan's third wireless carrier, J-Phone (since Vodafone KK and now Softbank) experimented with the first integrated camera into a phone. What the camera industry - also based in Japan - laughed at back then, rapidly spread and today over half of all phones in the world have cameras. In South Korea, for example all phones sold have cameras.

Therefore, when kids get their first cameraphone, it becomes an instrument of immediate gratification. A friend makes a face at MacDonald's, take a picture. The guy jumps with his skateboard, ask him to do it again and shoot it with the cameraphone taking a short video clip. Tomorrow the camera memory is full, delete these images and clips, and make new ones. Most of the images taken by kids on cameraphones are meant to be consumed within the next 48 hours, shown to a few friends and deleted. Almost none are meant to be kept.

D SMALLER TRIBES

For many generations teenagers have exhibited "herd behavior" i.e. they have copied each other. There has been an accepted style of music and preferred bands, as well as the way of dress, accepted brands of blue jeans or sneakers, etc. Today's teenager herd is splitting into smaller herds than ever before. Now kids will find their own groupings and dress by their own "tribe" such as those who are into skateboarding, or the Goths who tend to dress in black, etc. Each tribe has its clear culture of behavior, dress, music, drink, etc. Thus even a classification of teenagers as a group is becoming useless, the kids need to be divided into ever-smaller subgroups and microsegments.

Cellphone and identity

Kids are also associating own identity factors with the cellphone. As the cellphone is so critical to community contacts, it is even more important as a personal status symbol. For the youth, what used to be associations of being cool, adult, sophisticated, that were attributes strongly promoted for example to associate with cigarette smoking in the 1960s, and the youth strongly still associate with cars today, are increasingly associated with the cellphone. Or for our American readers, what for previous generations was associated in the first car of an older teenager, is now being associated with the right type of cellphone.

What phone one has, what covers on the phone, what ringing tone, what games, screen savers, etc., will all help the youth communicate with peers and others a sense of who that person wants to be. The 2006 survey by Mobile Youth revealed that the crossover happens around the 14th and 15th year. Up to age, 14 teenagers will accept their parents' choice of phone and the plan of the wireless carrier (mobile operator). However, from age 15 the teenagers insist on selecting the phone and service plan themselves, to suit their tastes and communication patterns with their friends.

Who may use the phone

Again, it is interesting to see how differently younger people use their phones with accepted inner circle friends, and others. For example parents, little and big brothers and sisters, favorite uncles, etc, will not be allowed to scroll through saved messages, etc. Even if it is a new phone, which the youth may be very eager to show off, such as its new camera feature or latest game, that youngster will not want the other to dig through messages and calling histories etc. to find out what personal communication that young person has been up to.

However, see the same youngster with his or her best friends at a McDonald's or Pizza Hut. They will all happily put their phones in the middle of the table, and all will reach out for each other's phones, play the latest games, look at the stored pictures, read through all latest messages, comment on them, etc

Sharing, connecting with community

Once a young person learns he/she is not alone, but can at all times contact friends, a much more interdependent relationship starts to build. Decisions are not made in isolation, even if late at night when a phone call would be frowned upon, young kids can contact each other silently via text messaging. Experiences are not felt alone. No, experiences are immediately shared, when they happen. In Finland Professor Timo Kopomaa was the first to document this in his groundbreaking book *The City in Your Pocket* in 2000 (translated from *Tietoyhteiskunnan Synty*, 1999). Professor Kopomaa highlighted how differently young people behave when they have the power of the community in their pocket:

> *Nothing is agreed upon or fixed in precise terms, the spectrum of individual choice is kept as broad as possible. A certain ex tempore lifestyle becomes more widespread. Both shared and private decisions are expected to be taken rapidly, and schedules are not determined precisely, because they can be adjusted along the way.*
> Timo Kopomaa, *The City in Your Pocket*, 2000

This generation is used to sharing experiences and to seek the opinions of friends, via the cellphone, at all times and in all situations. Moreover, not only knowing that the friends are there, always, this generation can now actively seek their opinions and support at any time.

E NATURAL BORN GAMERS

Another area where kids today behave differently is the virtual worlds of multiplayer gaming. It is not the first generation to grow up with videogaming. Those of us now in their forties had videogames at arcades, like Pong, Pac Man and Space Invaders. Those in their thirties grew up with home console games such as Super Mario Brothers and Donkey Kong and stand-alone PC games like Tetris and Doom. The twenty-somethings played with Playstations and PS2s, and Xboxes. But even those were mostly standalone or dual mode.

What makes Gen-C so potent in South Korea, is that they are so intimately familiar with the virtual worlds of multiplayer gaming. We will discuss virtual worlds and videogaming in their own chapters later in this book, so we will just summarize that the world's largest Massively Multiplayer Online Game (MMOG) is *Lineage II* of South Korea with twice the registered users worldwide than its nearest "Western" and better-known rival, *World of Warcraft*. In addition, of casual gaming, 25% of all South Koreans have played the same online videogame (*Kart Rider*).

Born to Share

Gen-C will be digitally literate and have multiple technical means to capture and generate content for themselves. They are also very apt at sharing, so they will borrow the friend's scanner, the big brother's digital camera and the uncle's faster computer or the aunt's CD burning drive to achieve what they want to do. They will also make do with their own equipment to a surprising degree. Even if their own cameraphone has very limited resolution, they will be frequently using it to capture significant events and memories, such as the big snowboarding meet, or the birthday party at Pizza Hut, or the partying that happened with graduation, etc.

> **15% of South Koreans play videogames on their cellphones every day**
> Source: Korea Agency for Digital Opportunity & Promotion 2005

Gen-C is also very used to sharing, especially using various peer-to-peer networks and social networking sites. In South Korea, of course practically all teenagers will have profiles and regularly visit friends inside Cyworld, which we discuss in the virtual chapter later. Teenagers make good use of digital memories, so a grainy image of a friend very drunk on a party a few years ago, may suddenly emerge as part of a birthday card, etc. Gen-C is very aware of the costs of sharing, so they will happily store digital memories on cellphones, iPods, personal computers, network hard drives, etc., and share them when they can do it at no cost, rather than use some of the expensive networking technologies. Then, again, Gen-C tend to be young, and thus arguably "irresponsible" with their money. They can easily spend all of the disposable income on the virtual world and feel no remorse about it.

Not like their parents

Members of Gen-C relate to cellphones and in fact, all networks in a different way to previous generations. The network is there not to interrupt me; it is there to serve me. The network will not control me; I control the interaction on the network. Let us start with the ringing of the phone.

For the youth today, it is also understood that not every call has to be answered. In Japan the young keitai (cellphone) generation has already developed a contacting ritual. Professor Mizuko Ito has observed that for the youth today, it is common to send text messages before phone calls, to "schedule" the time for a voice call.

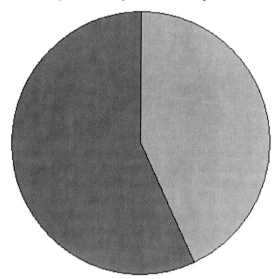

43% of South Koreans use Cyworld

Over 21 million South Koreans, 43% of the total population already maintain personal profiles in Cyworld in 2006

Source: SK Communications 2006

Generate own content

Gen-C used to creating own content using various digital means, both by sampling and copying from existing sources, and using the wide range of digital cameras, scanners, built-in microphones, etc. The generation is also very familiar with editing tools on the computer and will easily use these to edit, improve and crop content to suit their tastes.

My phone my money

For the Gen-C, the first financial instrument is not the credit card, a checking account or a debit card with a bank. It is the cellphone. This generation assumes it can pay for anything with the cellphone. In fact, where the phone account is a "post-pay" or contract account, it is usually the first tool of credit that Gen-C is exposed to. They may make a hamburger payment on the cellphone simply because their cash is low, and they know the phone bill will not become due until the end of the month. Very short-term credit. And once the carriers (network operators) enable it, kids will borrow from and lend each other money via cellphones, like invented in the Philippines, but launched of course in most Asian markets and now for example in Spain by Telefonica.

F SIMULTANEOUS PARALLEL NETWORKS

A key element of the currently connected society, is that its participants have learned to navigate multiple partially overlapping networks. The current digitally aware society is not easily manipulated. One wants a direct call, another hates the call but prefers listening to voice mail, a third will want the contact via e-mail, a fourth via SMS text messaging. Even more, we may have different preferences by the time of day, or type of communication, or the person contacting us.

Multiple connections, multiple communities

Similar to the multiple connectednesses with networks, we also have grown very adept at maintaining multiple communities. We have our family, our work colleagues, but also friends sharing a hobby, like buddies with whom to go watch football, or play golf, or bowling or play poker or bridge or whatever. We will very likely assign different communities amidst our work colleagues, especially if working for a larger organization. Before the cellphone, we did not have a strong selection of options to differentiate between our communities. Now we can even program our phone to ring in different ways depending on which person calls, etc.

Cool is digital cool

South Korea may seem distant to Western readers, but within Asia, South Korea has rapidly gained a status of being cool. Like perhaps what the UK or Italy may be for Europe, or the West Coast for America. In Asia, the concept "cool" is almost synonymous with "digital cool." In China, Japan, Malaysia etc South Korean popular culture from clothing fashions to hairdressing shops to pop music artists to soap operas. The pop singer BOA for example is regularly on the top of Japanese pop music charts. Korea is driving fashions and trends of what is cool in Asia.

The Asian countries value South Korean leadership very highly, in particular with youth-oriented areas, gaming, music, TV, fashions, etc and South Korean brands and concepts have expanded strongly and are now copied locally. Known as the Hallyu – The Korean wave of pop music, fashion and culture others are massive exports that add to cool Korea Youth culture. Even in London, Korean restaurants are now gaining the chic of being the new and trendy. Asian students in London from Philippines, Malaysia, Japan, China and Singapore queue up and pack out the Korean restaurants to experience sometimes their first opportunity of Korean community

Because the language is different (Korean is not spoken in any other country, and its alphabet is no that used in its neighbors Japan and China) many South Korean innovations have remained nearly forgotten within its own borders, often for years. For example the recent internet innovation of blogging. In 2006 about 8% of all internet users maintained blogsites. This was up from less than 0.2% in 2004. However, back in 2001, in South Korea blogging was so big, that a movie was inspired by blogging, in the hit film *"My Sassy Girl."*

The leading country, its pioneering users

We wanted to start with a chapter on the youth, because we both feel that the changes in behavior we see in the youth of today, as they relate to high technology, will migrate with them as they grow up, get their first jobs, and bring this behavior to the workplace.

This chapter summarizes lessons from various youth studies from Finland, Japan, Belgium, the UK etc, as well as South Korea. We can see a distinct form of behavior emerging with the youth, radically different from that of their elders. While multiplayer gaming, broadband internet, digital TV, portable music etc are all relevant innovations to the youth, the most empowering technology to them is their cellphone. For any impact that we may see of cellphones in the general public, consider the findings of this chapter. Generation-C, how radically will that behavior of the Community Generation alter the workplace and general commerce, media, government etc. when this generation graduates from college and starts to work.

Where the rest of the book is about areas where South Korea has a leadership position in the digital information society, this chapter looks at global trends in youth behavior. We suggest you consider this as the most forward-pointing chapter of a forward-pointing book. All of the following chapters show likely visions of a future for their respective areas whether in the home, in music, TV, gaming, robotics, etc. However, this chapter is a look into the soul of the consumer of tomorrow - and as far as we've found so far, all of the relevant research into the youth specifically in South Korea.

Case Study 1
Cyworld

When IT experts in the the Western world think of advanced social networking sites in late 2006, they tend to mention the video sharing site YouTube, picture sharing site Flickr, personal profiling and chat site MySpace, blogging, virtual reality worlds like children's Habbo Hotel or the service more for adults, Second Life. Often commercial social sites like book seller Amazon (with its user-reviews) or eBay (auctions, a social activity) are also mentioned. While these are among the concept leaders in the Western world for these given sub-categories, the world's most advanced social networking site, by a wide margin is Cyworld in South Korea. It already offers videosharing like YouTube, picture sharing like Flickr, personal profiles and chat like MySpace, blogs, virtual reality worls like Habbo Hotel and Second Life, plus commerce like Amazon and eBay - all rolled into one. Where Western single-purpose sites like YouTube, Myspace etc are used typically by about 20% of the total population, Cyworld is already used by 43% of the South Korean population.

Cyworld launched in 1999 in South Korea aiming for the youth segment as a personal profiling, chat and virtual site, and was bought by SK Communications (part of the SK Group perhaps best known for SK Telecoms). Cyworld was initially just on the internet, but its mobile arm was launched in 2003. Today as 100% of South Korean internet access is broadband, so too is all of the internet-side customer traffic for Cyworld. Of the mobile access, by far the majority is from advanced 3G cellphones.

As Cyworld is a multiple function online social site, it is almost impossible to describe all the features it has. We discuss Cyworld's impact to several related industries like music, gaming, virtual worlds, etc, in their own chapters in our book. But usually the most striking first impression of Cyworld to Western visitors is its user interface. The "minihomepy" (mini home page, perhaps blogsite would be best Western equivalent) with its "miniroom" (my room, similar to a Habbo room) and minime (my avatar, a digital puppet as my digital visual representative). These have a distinct "Asian" almost "Hello Kitty" type of appearance which can be thought of as child-like. These appearance factors may put Cyworld off especially to older executives in telecoms, IT and media. But don't be fooled, the Finnish variant has over 8 million subscribers in mostly Western countries, over a million in the UK for example. Don't be put off by the appearance.

The miniroom is your virtual property, like an island or house in Second Life or a room in Habbo Hotel. You can invite friends to visit your miniroom and engage with them there. You can entertain your friends for example share music and pictures there. Necessary for your room is your minime, your avatar. Like yourself, you will want to customize the minime. What dress, what haircut, is it male or female, etc. The same goes for the room, do you want a sofa, a table, a window, etc. Some of the basic customization is available for free, the rest of the customization costs.

Payment in Cyworld is handles with Dotori (the acorn) and these can be purchased for example by your mobile phone account. Then digital items inside Cyworld, whether user-generated content or branded content, can be bought and sold. When 43% of the total population lives inside a virtual world, then also every major consumer brand will want to be there. Cyworld had already 30,000 corporate/business customers with a digital presense inside Cyworld, selling over half a million items of branded content, by the summer of 2006. This is a healthy revenue stream for SK Telecom which makes a 40% revenue share cut on any digital content sold inside Cyworld. Cyworld earns 450,000 dollars per day out of its earnings of branded digital content sold inside the social networking site.

How big is it? 90% of all pictures sent from cellphones in South Korea do not go to other cellphones, they go to the picture sharing service at Cyworld. Music? In a country one sixth the size of the USA, Cyworld has already become the world's second largest digital music store behind iTunes by 2006, and notice while iTunes sells in dozens of countries, Cyworld sold music only in South Korea. As of YouTube? SK Communictions CEO Hyun-Oh Yoo pointed out that at 100,000 daily uploads, Cyworld generates more video uploads than YouTube.

Chapter III
Virtual Worlds

New Posts
* body picnic
* Just us girls
* john's house
* wheeeeeeee

Recent Photos

Miniroom: **Penthouse Suite**

Comments (7) View

Neighbor Comments
Hey! I love your room, but you need company! I'm on my way, neighbor.

Our Second Lives

Image courtesy Cyworld and SK Communications

> *Cyworld is like MySpace two years into the future."*
> **American users of both MySpace and Cyworld,**
> quoted in the *International Herald Tribune*

III
Virtual Worlds
Our Second Lives

A recent survey of the most desirable cars in South Korea revealed that the most popular car was the Porsche Carrera GT. That is not surprising. But the second-most desired car was "Solid Pro" a virtual car from the online videogame Kart Rider. Solid Pro finished ahead of Ferraris, Lamborghinis, Lotuses, Maseratis, Aston Martins, Mercedes Benzes and BMWs. Kart Rider is a videogame. A massively multiplayer online game that is sweeping South Korea and has become the most popular multiplayer videogame by number gamers. 12 million drive *Kart Rider* in South Korea alone - yes 25% of the total population. That is what becomes possible when we enter virtual worlds in the world's most digitally connected country.

A GROWING UP WITH VIRTUAL REALITY

Many know that South Korea leads the world in adoption of videogaming. Just about every statistic is bewildering. Yes, 25% of the total population drives imaginary cars in *Kart Rider*. 15% of South Koreans play videogames every day - *on their phones*. We will look at some of the more popular videogames in its own chapter later in the book. Gaming is only the first step into a deeper world where digital meets reality, in what we call virtual worlds. In the Western world, many are familiar with Second Life and Habbo

Hotel as typical virtual worlds, where humans create virtual personas as avatars and meet other people, interact, chat, collaborate, engage in trade and commerce, etc. This chapter looks at the virtual worlds in South Korea. Let us start with our virtual persona.

> **20% of South Korean cellphone owners use internet search on their cellphones**
> Source: NIDA September 2005

Me and my avatar

Recently South Korean culture has changed drastically with the growing popularity of avatars. Avatars are "video game like" cartoon representations - digital puppets if you will - of people that are used in virtual chat worlds and environments, and on mobile phones as screen savers. Differing from how users are identified inside traditional chat programs where users have only text identification or perhaps a small picture of the face of the users, or cartoon representation. Avatars are actual "virtual robots" which are usually three-dimensional, have form, clothing, haircuts, etc. One of the most popular avatar sites in South Korea is Neowiz's "SayClub" that has over 20 million subscribers, which is over 40% of the total population of Korea.

SayClub

In SayClub, the avatar initially comes only with underwear. The user has to then outfit the avatar to fit the kind of persona that user wants his/her avatar to reflect in its virtual world. Some want their avatars to resemble their real world appearance. However, more often the avatar can gain attributes - a dark haired person to be blonde for example or an overweight person to be slim - and of course wear clothes that the real user might not be able to afford.

Each additional item of clothing or accessory needs to be purchased and then dressed upon the avatar. As these kinds of environments grow, they soon gain a vast range of content such as clothing from the major brands and up to premium fashion designers like Gucci. It is not uncommon for South Koreans heavily into virtual worlds to spend more on the accessories and clothing of their avatar than their real wardrobe. An estimate by Daewoo Securities on the value of the avatar market in South Korea was 114 million dollars in 2004, doubling each year since 2002.

Mini Me

A particular avatar is that inside Cyworld. We discussed Cyworld already in its case study in the Youth Chapter, so suffice it to say Cyworld is like the Western online social networking sites of Second Life, MySpace, Flickr, YouTube, eBay, Amazon, Habbo Hotel, and iTunes, all rolled into one. Cyworld exists on both broadband internet and 3G cellphone.

The avatars in Cyworld are called Mini Me. A small virtual representation, more like Habbo characters resembling tiny Lego people than the more humanoid avatars in Second Life, the Mini Me is nonetheless a vital element of the Cyworld experience. Like more advanced avatars like those on SayClub, also on Cyworld users can customize their Mini Me with standard features from Cyworld, and by paying more, buying branded and custom clothing and accessories.

When nearly half of a nation's population engages inside one virtual world, it also becomes a common place for all kinds of other human interaction. Dating in South Korea often starts inside virtual worlds, using avatars. The real people behind the avatars feel they can be more open inside the virtual worlds and interacting with each other virtually, before meeting in person. However, these relationships can well develop into real dating or real friendships.

Virtual playgrounds

When millions of uses join into one virtual world or multiplayer game, tens of thousands or even hundreds of thousands might be active at the same time. Virtual worlds become the modern equivalents of the village square, the school playground or the neighborhood pub and bar. People gather. Not only to participate in the game or activity, but also to "hang out" and meet with friends.

At the one extreme are MMOGs (Massively Multiplayer Online Games) or what are also often called MMORPGs (Massively Multiplayer Online Role Playing Games). In the Western world, these would include *Everquest*, *World of Warcraft* and *CounterStrike*. *World of Warcraft* is so far the biggest Western online gaming environment with 7 million gamers. The world's biggest MMOG, however, is the latest edition of the South Korean medieval fantasy game, *Lineage II* that has 14 million users worldwide and 7 million users in South Korea alone. Editions of *Lineage* exist in English, Chinese, Japanese and the MMOG has millions of users in each of those editions. We will discuss *Lineage II* in detail in the Case Study to this chapter, and also some details in the gaming chapter later in this book.

While digital environments like Second Life reach users measured in the millions, and offer opportunities for commerce, various major brands and corporations enter them to make business or provide their services also in the virtual world. Some like Habbo Hotel spread to over a dozen countries and already approach ten million users. Not gaming like the customers of *Warcraft*, *CounterStrike* and *Lineage*, but more simply "playing" and meeting friends, and even building virtual business enterprises such as setting up clubs or manufacturing digital content for the avatars or properties, etc, as we see starting to happen for example in Second Life.

B CYWORLD

The most advanced virtual environment, both by what all is possible in the world, and how far it has invaded the total real world, is of course, in South Korea. It is called Cyworld. When first visiting Cyworld many Western users may think that it is too "cute" for Western eyes, with a distinct *"Pokemon"* or *"Hello Kitty"* kind of "Asian" element of cute, simplistic, even childlike cartoon appearance. Don't be deceived, Cyworld is by far the most advanced virtual ecosystem and the most complete virtual economy, as well as the most complete social networking service yet created anywhere.

Miniroom

Like children setting up their own virtual rooms inside Habbo Hotel - or a simpler private domain than the homes and islands on Second Life, Cyworld also offers the chance for users to set up and decorate their room. This is the natural meeting place where friends can be invited. In addition, the Mini Me can go visit rooms of friends, or go to public rooms like virtual bars, clubs, shopping malls etc to meet friends or make new friends.

Not unlike dressing up your Mini Me, the Miniroom can be decorated with furniture and furnishings, just like your home or office, or it could be aspirational of what you would like your real life to be. So if you would like your room to be a clean white, minimalist, with some abstract art, and a grand piano. So be it. Another person may want to copy what the current real home is like.

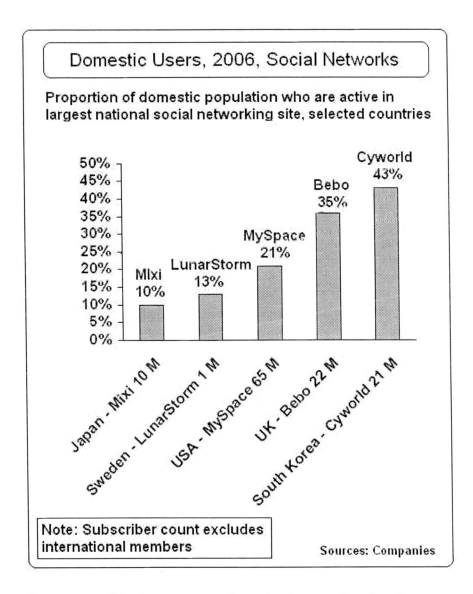

Someone may like the room to continuously change and evolve. Someone else likes bright colors, and so forth.

The Acorn, the business engine

Money makes the world go round. It has been a particular engine for propelling Cyworld into such success. The monetary unit inside Cyworld is the Dotori (acorn), which is worth about 10 cents. Cyworld content and properties such as decorations and furnishings for the Miniroom and customizing and updating the avatar are paid for in Dotori. Third party content is paid for in Dotori. In addition, Cyworld users can sell content they have bought or created, and charge for it in Dotori. Dotori are most easily purchased through the cellphone account. Right from its beginning Cyworld has been built around a robust commercial foundation allowing users and third party content owners to fully enjoy economic benefits of membership.

How much money is involved? The trade of virtual goods and services inside Cyworld was worth over 450,000 dollars per day by 2006, which works out to a turnover of over 10 million dollars worth of trade per

> **The measured highest broadband speeds in Canada in 2006 were 5 Mbit/s**
> Source: Analysys October 2006

month. It is a remarkably vibrant marketplace if you consider Cyworld was not set up explicitly as a trading community such as Amazon, eBay or iTunes. Moreover, while we mention iTunes, Cyworld is an exceptional music store.

Music world

Universal Music said at the iMobiCon digital convergence conference in Jeju South Korea in November 2005, that Cyworld had become the biggest single music outlet in all of South Korea, in effect the largest music superstore in the country.

If our readers are familiar with the economics of ringing tones, ringback tones and MP3 full track music downloads to cellphones, they might think that Cyworld sells these music services. That is not the case. Music sold inside Cyworld is consumed also inside Cyworld. The three biggest categories of music sales are background music for your Miniroom, the Welcoming song to your Miniroom, and songs as gifts to give to friends when you visit them.

As Cyworld users want to decorate their Minirooms, they will also want to be good hosts, and play music to friends who visit their rooms. Cyworld teamed up with Universal Music to offer its full catalog as MP3 tracks, which are played as background music inside the Minirooms. Soon Universal was selling 100,000 songs of full-track MP3 songs per and by 2006 all major record labels were active inside Cyworld with total daily sales exceeding 200,000 songs per day.

What is your Welcoming Song

As a particular innovation, the Miniroom also can include a Welcoming Song. Much like the ringing tone on the cellphone, users inside Cyworld can set a Welcoming Song to play for any guests when they enter the room. After

---

The measured highest broadband speeds in South Korea in 2006 were 18 Mbit/s
Source: Analysys October 2006

---

all, a good host will play music to friends. The Welcoming Songs can be selected from a vast catalog of current and past music, with full authority from the record labels and artists, with full royalties paid.

The recording industry loves this idea; after all if I select a given James Bond track as my Welcoming Song, odds are that I have already bought the same song on a CD and fully paid for it. Perhaps I have also bought the same song as a ringing tone and paid for it a second time. Now if I want the song as a ringback tone (waiting tone), I would be paying for the same song a third time. Welcoming songs allow me to paying for the same music yet again. And the gimmick? Much like the old juke box, every time the Welcoming Song is played, the owner of the Miniroom is charged per play, of about 40 cents. Every single time the Welcoming Song is played. No wonder the recording industry loves Cyworld.

C BLOGGING

Nevertheless, its not only a children's playground and music store. Cyworld gets really interesting when we add the blogging part. Blogs, or web logs, are now becoming familiar in the West on the internet, in what started as mostly young people posting personal diaries and thoughts. Today blogs are cutting edge news sources for the news media, for example with services such as Reuters offering blog content to support their breaking news coverage. While blogs hit the mainstream in the West around 2005, in South Korea, blogs have been commonplace and in the public domain for years. A movie was released in 2001 about a girl blogger, called *"My Sassy Girl."*

As Cyworld expanded its offering, it soon also became the most popular blogging service in South Korea. Now Cyworld jumps ahead of most of its Western social networking service rivals. Cyworld's Minihomepy (Mini Home Page) service has become the most popular blogging - and mobile blogging - service in South Korea. 42% of the total population has set up a personal profile at Cyworld, and most users also actively maintain it, hence blogging.

Be my friend?

Part of Cyworld's blog service is the ability to give visitors a classification, either public or friend. In this way users can designate close friends to have access to all private information and for example a teenager can exclude the parents, giving them only normal visitor rights, and then giving their best friends the full access to all more personal and intimate thoughts, blog entries, images etc. The social side of Cyworld is particularly strong and is reflected in its roots: the original concept was an online dating service, as Cyworld founder Young Joon Hyung explained in 2006 in the Plus Eight Star (+8*) White Paper on Cyworld:

> *At the beginning, I was developing a dating service similar to 'match.com' but I realized the Korean society would require a stronger online certification process to enable enough trust to make the service successful. In addition, I wanted to create a widely open public meeting place. Then I got a quick idea from 'sixdegrees.com' and its concept that every individual is connected to another through a maximum of 6 degrees of separation. In my PhD research, I was working on how to adapt the idea of Enterprise Resource Planning (ERP) to individuals in order to reach the full potential of social networks in a fully sociable cyberspace. I labeled this concept*

Chapter 3 - Virtual Worlds

'Personal Resource Planner' (PRP) and implemented part of it in Cyworld.
- Young Joon Hyung, Founder of Cyworld, quoted in Plus 8 Star *White Paper on Cyworld*

Differing from traditional text-based blogs, Cyworld today combines social networking with blogs and offers it all in full multimedia. This again is

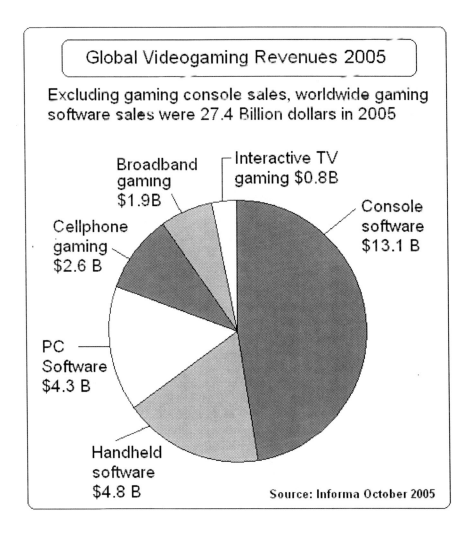

another example of how demanding the South Korean environment has become when practically all cellphone users have cameraphones and over half of phones are 3G high-speed phones. The customers expect picture, videos and sounds in social networking sites. Today in Cyworld through the Mobile Cyworld users can rapidly upload blog text, pictures, video clips, etc and post them to the Minihomepy and share with friends. It is no longer a new feature to enable direct uploads from cellphone to blogsite, rather it is already expected, taken for granted. 90% of all pictures transmitted from cameraphones in South Korea, end up on Cyworld's picture sharing sites and personal blogs.

Moblogging

Cyworld's blogsite is provided by the internet arm of SK, South Korea's largest telecoms provider, and there is direct mobile blogging opportunity from the mobile arm of SK. The traffic in pictures and blogging posts is

---

10% of British population upload pictures from cameraphones so social networking sites
Source: Telephia January 2007

---

enormous, with picture posting and viewing of picture galleries considered now the most addictive part of Cyworld. Mobile blogging (mostly for Cyworld) is the second biggest value-add service in Korea today, ahead of music, behind only gaming revenues, averaging 3.40 dollars per user per month.

Cyworld in context

The Western media has been impressed with the rapid rise of MySpace, to over 100 million users. MySpace is strongly based in the USA, with for example the British using their domestic Bebo service, the Japanese preferring their Mixi, and so forth. In the USA, over one in four Americans has set up an account in MySpace. While Cyworld has "only" 21 million users in South Korea, within the South Korean population, that means 42% of the total population - and the user numbers are still growing. For the adoption of a social networking site in any country, Cyworld is the clear global leader.

An interesting comparison can now be made in America. Helio the MVNO (Mobile Virtual Network Operator) in America launched Cyworld into America in 2006. Early users familiar with both MySpace and Cyworld were reported in the *International Herald Tribune* of 25 September 2006, saying once they started to use Cyworld, they barely returned to MySpace, and that Cyworld was like MySpace but two years into the future. We would put it this way. Cyworld is the world's most complete online virtual world and social networking site. MySpace was opportunism, being in the biggest online market, USA, at the right time.

D FRIENDSHIPS AND BUSINESS

Many Western observers have been amazed at how social networking sites

> ## 30% of South Korean population upload pictures from cameraphones to social networking sites
> Source: Seoul Magazine December 2005

deal with friendships. Teenagers in America measure their popularity by how many friends they have on MySpace, to the degree that now Greenpeace and other such organizations have invited MySpace members to become their "friends". This is similar to the professional sites like LinkedIn where many race in contest to have the biggest personal networks. Again, this is nothing new. Forming friendships and then building human relationships is a core element of Cyworld. Like American teenagers in MySpace, South Korean teenagers measure how popular they are by how many friends link to their Minihomepy's.

Cyworld facilitates the forming of new friendships through its "Becoming Buddies" feature, which is a creative human relationship management system. Becoming Buddies allows sharing of information, pictures, blogs etc as well as bonding in friendship.

30,000 Businesses in Cyworld

However, even more, once you have two out of every five members of the population in your virtual world, the whole economy takes notice. Every brand and company wants a presence inside Cyworld. Today all significant South Korean businesses already maintain a presence inside Cyworld. It is no longer a question of "should" Coca Cola or Nike or Ford find marketing tools to join MySpace. In Korea, every consumer brand has to be inside Cyworld. 30,000 businesses, offer over 500,000 items of digital content for sale already. An honest virtual economy eco-system.

The Emotional Messenger - Nateon

Among the many supplementary services on Cyworld, one of the most advanced is Nateon the emotional messenger. Most instant messenger services and chat rooms today have poor facilities for communicating emotions and feelings. Here is again where South Korea has developed a lead, in part because so many people are there already, and in part because so many users spend so much time in Cyworld. Emotional Messenger was launched in January 2003 and rapidly overtook MSN Messenger and became the most used messenger service in South Korea due to its "emotion code".

As it matured as a service Nateon adopted ever more typical virtual versions of South Korean behavior. It offers gifts and allows users also to make gift requests. It allows four-way video calling, as well as the obvious music album.

World Invasion

Cyworld started its global expansion with China where it has passed a million users, and now has already launched also in Japan, several other Asian markets, USA (on the Helio network) and the first European launch country was Germany in 2006. Cyworld hopes it will mimic the success of other South Korean digital innovations such as Ringback (i.e. Waiting) Tones and MP3 files sold direct to mobile phones, and digital TV broadcasts to mobile phones.

E LEARNING THROUGH VIRTUALITY

Thirty years ago training simulators were complex multi-million dollar custom-devices run by mainframe computers, to help train airline pilots, submarine captains and crews of main battle tanks. As computers became cheaper with the PC, and applications appeared that allowed rapid development and modification of calculations, the simulations entered the high risk and high cost area of investment banking in the 1980s. By the middle of the last decade PC based simulators had spread as far as online gaming for example in various virtual team gaming, such as fantasy football leagues, baseball leagues, hockey leagues etc.

> **Pearl - Learn to Dance on your phone.** If you want to learn the latest dance moves and are a bit shy about your dance ability, no problem. Close the door, call up the karaoke, select the song, set the tempo to slow and turn on the dance tutor. A stick figure will patiently show you all the steps, without ever mocking your performance, until you know the moves, and adjust the speed to normal.

Simulations

With simulations and the ease of use of spreadsheet calculations, many very complex management decisions and their impacts could be modeled rather easily. The advanced mathematical modeling discipline of econometrics combined with ever more powerful computers made it easy for the development of almost any kind of business simulations and soon thereafter, the various scenario development in social, political, educational etc areas. What is important to note, is that until the last decade, the use of simulations was something you went to university to study, and needed custom software applications to utilize.

Games and learning

Playstation games such as *Grand Theft Auto and Flight Simulator* changed all that. Young people of today know that almost any activity of their life can be simulated, and most simulations are available online, often on their personal cellphones. In Finland there is even a cellphone based virtual tutor/mentor for engineers called *Metronerd,* to learn how to interact with "non-engineers" in to prepare engineers for dating situations. The current generation in high school is also that generation which remembers fondly the toy known as the tamagotchi. This is the first generation to have had a virtual pet before ever knowing a real live pet.

The same kind of techniques are used in the virtual boyfriend/girlfriend, also on the cellphone, to learn the basics of dating before the young teenager starts off with the first real dating experience. Growing up today is very competitive and causes a lot of distress for youngsters in any country. In South Korea on their own cellphones, teenagers know they have every tutor and mentor they would want. Feel like brushing up on English skills or preparing for the math exams? No problem, there is the training available on the phone. Nobody need to know that you take tutoring.

25% of South Korean population have played the same multiplayer online game, *Kart Rider*
Source: Nexon 2006

Fatigue and stress

As South Korean youth and young adults spend so much time in videogame environments, on virtual worlds, surfing the web and chatting and messengering on their phones, PDAs and laptops, there also sets in a great degree of digital fatigue and various aches and pains from repetitive stress. South Korean society is now adjusting for these symptoms. For example the gaming characters inside South Korean games will become virtually fatigued after playing too long - your character is no longer as capable as normal. This is a clear sign to the gamer to quit and take a real break in real time. Too much gaming and play inside virtual worlds is physically too demanding.

Growing up virtual

While virtual worlds allow easy forming of new friendships, they also tend to make it faster and easier to drop friendships with little or no notice. The commitment to virtual friends seems to have less depth and sustainability. Certainly the member who has 10,000 friends is unable to neither recall all of them nor maintain any kind of personal contact with all. Nevertheless, even in closer friendships, one side of the friendship can be very hurt when the other party almost randomly seems to cut off the contact and ignore the previous friend. We are too early to make definitive judgments or draw conclusions. It has also been argued that as we become increasingly urban as a society and as education and employment often take us to different cities and even countries, an avatar-based social connection in a virtual world may be much stronger than the occasional letter, e-mail or phone call.

Finally there is the concept of employment inside virtual worlds and MMOG types of massively multiplayer games. In addition to the superstar gamers who can attract sponsorship because of their mastering of a given game, a new full-time career opportunity of significant earning potential has emerged inside gaming. It is called Farming for Videogame Gold. These gold farmers spend 8 hours a day, five days a week, traveling to the dozens of known locations inside the game where various treasures appear or grow, such as ammunition for the weapons, food and health for sustaining, and spells etc that may be of value for gamers. These farmers collect the various consumables from the game, and then go to online trading services like eBay and sell them for real cash. We will explain gold farmers in more detail in the Gaming chapter later in this book.

Case Study 2
Lineage II

For some years virtual worlds online, mostly in the form of MMOGs, Massively Multiplayer Online Games, were a murky hidden area of geeky gamers. Early MMOGSs like *Everquest* and *Ultima Online* grew to a million users. Today the best known MMOG in the West is *World of Warcraft* with seven million users, but the biggest MMOG in the world however, is *Lineage II* of South Korea with 14 million customers. The game was developed by NC Soft.

A typical role-playing multiplayer environment set in the Middle Ages, *Lineage* is based on a comic book of the same name by Shin Il-sook, and launched as a multiplayer game in 1998. The original *Lineage* reached 4 million gamers (then the world record) and *Lineage was* into its second edition as *Lineage II* subtitled *The Chaotic Chronicle*; the latest edition is *The Chaotic Throne*. The game resembles previous games such as Dungeons and Dragons and the gamers can appear as wizzards, elfs, knights, princes, etc. The game has Korean, Japanese, Chinese, English and Thai language versions.

Gamers select a character type (elves, princes/princesses, knights, wizzards and dark elves) and have typical "*Dungeons and Dragons*" types of skills and abilities such as spells and magic, while killing monsters and seeking treasures. There are also virtual pets inside Lineage. Very relevant to the *Lineage* experience is fighting between gamers.

Gamers who come from the console gaming world (Playstation 2, Xbox etc) and on stand-alone games on PCs feel that online gaming with real people is a whole new dimension to gaming, after which console games seem no longer challenging enough. Similarly gamers who play smaller online games or are active in virtual worlds, once when they "cross over to the dark side" or join MMOGs, find that lesser games and environments no longer satisfy their gaming needs in the same way.

> The gamers in *Lineage* spend weeks developing skills and then months exploring their abilities and playing inside the game. As there is no logical linear "task" to complete like in simpler games, the world and experiences inside an MMOG will be different for every gamer, and also different for the gamer at different times.
>
> It is not surprising therefore, that gamers easily spend hours per day inside *Lineage* and do fall asleep at the screen out of sheer fatigue.

NCsoft, Lineage, The Chaotic Chronicle, The Chaotic Throne, the interlocking NC logo and all associated logos and designs are trademarks or registered trademarks of NCsoft Corporation.

Chapter IV
Intelligent Home

Digital living

Image courtesy *IT Korea Journal*

> *"When I examine myself and my methods of thought, I come to the conclusion that the gift of fantasy has meant more to me than any talent for abstract, positive thinking."*
> **Albert Einstein**

IV
Intelligent Home
Digital living

When advanced technologies first appear, they tend to be in military and government uses. Next they arrive into business and educational uses. When they come to the consumer they tend to be in individual devices that are bought. However, when homes are built with the technologies in them, which is when truly life has fully embraced those technologies. Computers were first used in military to crack enemy codes in the Second World War. Eventually computers arrived to business use and then they even came to our homes as stand-alone PCs, and built into our cars, Playstations and cellphones. Now the home itself is becoming intelligent, as are our clothes and the very life all around us. Let's see how far digital living has come in the most wired country, South Korea.

A INTELLIGENT HOME

Starting with the home. Many futurists have promised us the intelligent fridge, which knows when we are running out of milk and orders it from the store. Such innovations were described already in Hannula and Linturi's book *100 Phenomena* in 1998. Today most home appliance makers have exhibited smart devices from the intelligent washing machine to the teapot that can be remotely activated via SMS. In South Korea the concept of the intelligent home is moving to the next stage, connecting the various appliances and enabling them to communicate with each other as well as the homeowner and his/her family.

Security

Wireless security systems have been around for business/enterprise customers for some time, and some wealthy homeowners have also installed home security systems in all parts of the world. This area of the intelligent home has developed so far in South Korea that today kits of home security systems can be installed by the homeowner, and rather than going via some third party solutions, the security system calls - and sends video - to the owner of the home and his/her cellphone. One such system was described by *IT Korea Journal* in the summer of 2005:

> *B-kyung System showed wireless home security system 'Izen', which is a burglar alarm device capable of arming, releasing and emergency calls conveniently with a remote controller at home. It is a high-tech device that protects home from intruders and can be easily installed at home by users.*
> Source: *IT Korea Journal* July-August 2005

While home security is one obvious area of the intelligent home, another rapidly evolving one is that of home entertainment. Current South Korean services and devices recognize the multiplatform nature of the consumption of the services. Companies such as Samsung, SK Telecom, Commax and LG each offer converged entertainment solutions which allow consumption of music, photos, TV and video as well as videogaming, across multiple devices and technologies including digital TV, broadband internet and cellphone, and across a wide range of wired and wireless technologies including 3G, WiBro, DMB.

Remote control

Each of the South Korean cellular telecoms providers offer the remote control of home utility services such as controlling lighting, gas and heating remotely via cellphone. For example SK Telecom is currently selling a service to its subscribers that allows people to leave video messages if the owner is not at home. When you ring the doorbell, if the person is not at home, the connected video camera offers the opportunity to send a recorded message to the owner, delivered then by SK Telecom to the owner's 3G cellphone. Again to show the user-friendliness of how far this service has evolved from the science fiction to truly useful - the owner of the home can then press a button on the phone, and unlock the front door to allow the guest into the home. Not only is it convenient for the visitor to contact the homeowner who happens to be absent, it also allows the homeowner to be a good host, and invite a friend into the home, as the homeowner will then hurry home.

> **Pearl - Remote control of home via cellphone.** Today several home security and management systems in South Korea offer remote control via cellphone. Think you forgot to turn off the stove? Now no need to drive back home, just use your cellphone and make sure it is off.

What of the pet?

In the country that first introduced the inter-species cellphone to human translator (dog barking converted to SMS text messages), now SK Telecom in South Korea has launched the full pet maintenance system that is controlled via cellphone. The owner can feed the pet(s), watch the pets via video, and do remote medical diagnosis, all remotely via a 3G cellphone.

And home wiring

Another area of big development is utilizing the electrical wiring of the home, with PLC (PowerLine Control) devices and systems connected to it. KyungDong Network for example offers a full home control system that can easily be retrofitted to existing houses, using PLC and thus requiring no additional wiring. The system has a home controller, which can manipulate not only home lighting and heating but also the cooking and heating gas valves, door locks and alarm systems.

Other home control systems are built on RF (Radio Frequency) communication, such as that by IK Tech which mixes internet application protocols and RF technologies to allow remote control of the home systems via the web as well as at home from a central controller. Other solutions exist

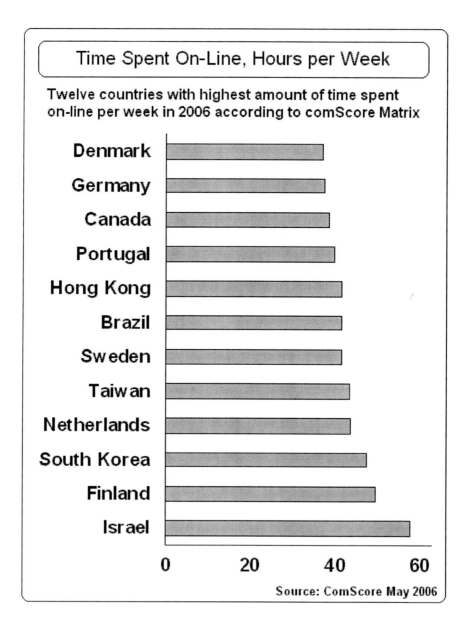

utilizing for example Bluetooth, SMS text messaging, WiFi and other technologies.

Intelligent door locks

South Korea is also moving rapidly into the era of the cellphone being the full digital key in addition to being the digital wallet. InnoAce has introduced a digital door lock for homes and offices, which has a cellular telecoms module, allowing users to open home door locks digitally with the cellphone. As can be imagined, such a system is not limited to door locks and will then naturally include modules for lighting, heating and air conditioning controls, as well as fire alarms, remote cameras, motion sensors and window locks. The cellphone is becoming the ultimate remote control. Another variant for controlling locks and replacing keys is to do intelligent locks via voice pattern recognition, iris scans and other biometric sensors such as face scans. Hyundai has introduced several such technologies for the intelligent home in South Korea.

Is commercial reality today

One sign of how far South Korean intelligent home concepts have evolved is the extent to which leading South Korean suppliers have branded their intelligent home offerings. Samsung calls its line Homevita, LG offers HomeNet products. Seoul Commutech's brand is Ezon. KT brands its intelligent home services as HomeN, and SK sells its offering under the Homecare brand. Obviously these companies have moved well past the piloting stage to selling commercially ranges of products and services.

B INTELLIGENT CLOTHING

The fashion world was taken by storm in the summer of 2006 when the first intelligent blue jeans were introduced in France by the premium brand Uranium Jeans. The jeans have a flexible text display patch at the back where users can send text messages. While exclusive fashion brands are now starting to launch digital clothes and fashion accessories, the South Korean IT industry is hard at work to introduce a wide range of wearable computing and communication gear.

Some of the most obvious gear has been demonstrated by numerous suppliers around the world, such as the eyeglasses/goggles that project an image so looking through the glasses gives the same visual experience as watching a giant screen display. However, on the side of the MP3 player, miniaturization has now been continuing and now it is possible to make the

MP3 player so small, that it can be fitted into an earring in the form of a cube.

Dispersed computing

Dispersed computing is coming, again not exclusive to South Korea. The individual elements of most parts of a personal computer, except for the display, keyboard (and the big battery to mostly feed the big display) can be made to fit a cellphone. When dispersed into individual elements, it is possible to build a computer into for example a jacket where no part is particularly bulky and the total weight of the jacket is increased only by a matter of a hundred grams or so. If we assume eyewear or goggles to take care of the display (with a battery built into the frame of the eyeglasses), it leaves us with the keyboard.

Some solutions have been already offered for keyboards from roll-away keyboards, flexible keyboards and projection keyboards, but perhaps a

---

19% of German cellphone owners send picture messages

Source: m:Metrics March 2006

---

more innovative adaptation is one using movement. When clothing is fitted with acceleration meters and velocity sensors, we can do without a keyboard or keypad for our data entry. Just moving our arm in the shape of a given letter can be programmed to respond to hitting that key on a keypad. Of course with this technology, if you want to type a message, you may end up looking like someone practicing for a fencing lesson, without the sword. Various game developers are very keen to deploy this kind of technology as has been seen on the Wii gaming console of Nintendo and Apple's upcoming iPhone has some of this technology apparently as well.

Wearable computing

The movement towards wearable computing is an inevitability. As the basic elements of the computer are continuously reduced in size, they also start to consume less electricity. We are seeing it in small steps today, from the

Bluetooth earpiece for our cellphones and the recently announced SMS text messaging wristwatches by Seiko and Citizen, and the controller units of the Wii gaming console. As modern smartphones are in fact palmtop computers, we are nearing the moment when computers truly will be worn. Then another possibility that emerges is then the solar cells that can be added to a jacket or overcoat, to recharge computer batteries from the light of the sun. These have already also been exhibited by South Korean suppliers.

Bridging the gap

As society interacts ever more in digital ways, we will then need now ways to bridge the gap between humans and machines. Mark Curtis explains in his book *Distraction*:

> We should design communications technologies, which allow us to communicate better. Half of good communication is listening – this

---

42% of South Korean cellphone owners send picture messages

Source: NIDA Sept 2005

> is often forgotten. Yet in small ways software can demonstrate listening of a kind: increasingly a standard of good design is that icons respond to the user when a cursor is rolled over them to demonstrate that they are active, or a suitable target for dropping another item. In a sense the computer is saying: "yes I know you are there and I'm giving you hints on what you can do". Another tiny example is the tick on Nokia phones which tells you that a task such as saving a telephone number has been completed. It's the digital equivalent of a nod. Digital body language can be even more explicit.
>
> Mark Curtis *Distraction* Futuretext 2005

We will explore the clash of humans and machines in the robots chapter later in this book, but these issues have a significant bearing in the way machines literally embrace us in wearable computing.

Digital Monitoring

Another area is developing the ability to read minute changes in human electrical charges. Devices can be embedded to read data while shaking hands for example. South Korean IT developers are working to bring the early prototypes into these kinds of communication tools to diminish our need to carry cumbersome equipment upon us to exchange information.

A little bit further down the line in time is the "bio shirt", an intelligent shirt or blouse that will monitor the health of its user. The IT Korea Journal explains:

> *The bio shirt is an innovative, ubiquitous healthcare application, which continuously monitors the health status of its wearer. The shirt with an integrated sensor and a device collecting physiological signals and transmitting them to the healthcare center are the two main components of this wondrous system. With this revolutionary apparatus, we can now receive quick help in the case of an emergency, and our body will be routinely checked for any alarming signs.*
>
> IT Korea Journal September-October 2005

Moreover, the R&D in South Korea is already working hard at what comes next. One is the use of color codes that are printed on clothing. These can be read by machine to act as our ID Cards. When combined with online purchases, we could for example walk into the airport and onto the airplane without a passport or ticket or even cellphone, as all the necessary data is linked to our ID, which we wear on our shirt, and is machine-readable on color codes.

C NEAR THE HOME

Moving from the home, several areas of living are also being now upgraded for digital connectivity. Our neighborhood supermarkets accept mobile payment, coffee shops have WiFi internet access and even the bus stop is becoming intelligent. We examine various shopping innovations in the Online Shopping Chapter, but lets examine that intelligent bus stop a bit more here

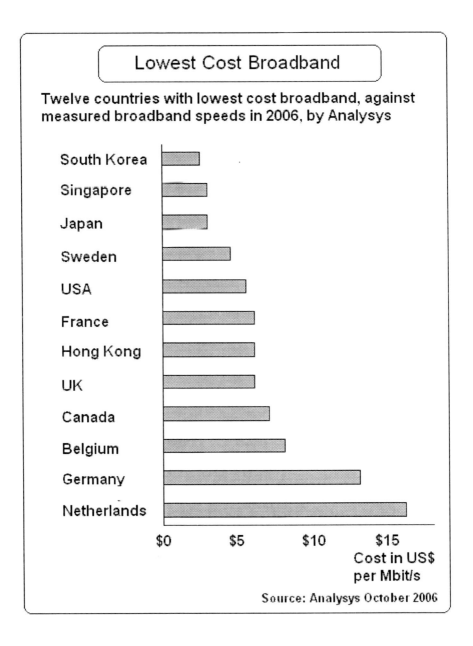

Intelligent bus stop

Another daily location for most commuters is the bus stop (tram stop, subway stop, train platform). KT has been working on the intelligent bus stop concept to develop it beyond the mere stopping station on a route.

We have seen train stops and increasingly bus stops with real-time bus information in many countries. Now KT offers the chance to send the real-time information to bus travelers' cellphones to any place at any time as requested by the traveller. The current South Korean intelligent bus stops typically already feature such features as time display and next bus display. And advertising at bus stops tend to feature 2D Barcoded information links to get instantly more information or advertising to users without having to type any further information on the keypad. We will discuss 2D Barcodes more in the Shopping chapter later in this book.

Ubiquitous Dream Hall

The dreams of the future merge in the vision of the Ubiquitous Dream Hall. The South Korean Ministry of Information and Communication (MIC) has set up a near-future vision of the intelligent home at their corporate HQ in Seoul and they call it the Ubiquitous Dream Hall. The Ubiquitous Dream Hall contains many futuristic devices already commercial available in South Korea as well as prototypes of near-future systems that will be launched soon.

The rooms in the apartments have latest home media devices, kitchen appliances, and robots which all are interconnected and can be controlled remotely. The intelligent wardrobe knows your selection of clothes and helps decide what to wear by projecting information and suggestions onto the mirror according to selected colors and the prevailing weather conditions. We discuss it in the case study to this chapter.

Large flat screen monitors and TVs throughout the Ubiquitous Dream Hall are used to project the day's agenda, traffic information, news, messages, etc. Multiple screen views on large screen displays allow multitasking such as catching up on the morning TV news while checking the weather and traffic reports online.

Robots do the cleaning while the whole home connects with the car telematics to update such information as a change in schedule and how that impacts on the optimal route based on today's real-time traffic congestion. The Ubiquitous Dream Hall is an evolving experience, constantly updated to feature the latest devices and services.

Extends beyond the home

Another common feature of living in Digital Korea is citizen bulletin boards, related to the various government departments that impact the lives of South Korean citizens. Each government department maintains bulletin boards where citizens can post comments. Seeing what other citizens have said helps engage the population. The South Korean council employees are encouraged to actively and rapidly post responses to citizen comments. South Korean citizens are very demanding which in turn makes the councils highly motivated to be responsive as these public bulletin boards are obviously visible to all.

> **14% of South Koreans check the weather report on their cellphones**
> Source NIDA September 2005

     A good example of how deeply the local government becomes involved with its citizens is adapting to personal "stamps" or personal seals used in Korean culture often as an alternate to a handwritten signature. The traditional stamps were wooden ink stamps with the personal seal written in Korean alphabet characters. Thus stamping a document was the equivalent to signing it. From 2002 electronic pay kiosks were introduced first in Seoul and then across the country, enabling citizens to access standard forms at all hours. These electronic forms were also very sophisticated allowing the electronic imprint of stamps (as important as signatures in Korea) as well as up to five levels of security protection. Since then video phones have been also included, to enable sign language for those with speech or hearing impediments to be able to digitally connect with council employees. We will examine the role of digitalization to the government in its own chapter later in this book.
     Road congestion systems and digital data transmissions allow for greater benefits to commuters. Real time traffic congestion data is now available to be accessed from cell phones at any time with further links to the nearest traffic monitoring cameras. These help drivers find the optimal routes on congested days. We look at the automobile and its digital evolution in the telematics chapter.

To finish living

Living in Digital Korea is becoming richer and more fulfilling in some ways due to ubiquitous connectivity but in other ways life is also becoming ever more stressful. When all drivers try to optimize their personal rush hour driving routes, it means that while bottlenecks perhaps ease up, the whole city eventually adjusts into a semi-permanent gridlock. A peaceful walk in the park on the weekend may be interrupted by our cellphone bringing images of our in-laws at our front door. And so forth. With the improvements come new problems. At least the IT industry in South Korea is hard at work at solving those as fast as they appear. Probably almost all who have new digital assistants in the home would not go back to life before them, such as Western adults would not want to go to times before the refrigerator, microwave oven and vacuum cleaner.

Case Study 3
The Intelligent Mirror

One of the latest gadgets exhibited by MIC at their Ubiquitous Dream Hall is the intelligent wardrobe. We all have wardrobes where we store our clothes. Many have a mirror on the doors. But what if rather than just a mirror, we combine a video camera, flat panel display some clever optical gimmicks, and a computer. Now we have the elements for the intelligent wardrobe, or perhaps, more descriptively the intelligent mirror.

Are your friends always complaining that you don't know how to dress? Do you hear that your shirt and tie don't match? Ever wonder which pair of slacks fits which sweater? No worries anymore, the intelligent mirror will tell you.

The Intelligent wardrobe uses the camera to scan all clothing it sees you wearing in front of it. As it learns of the actual appearance of your clothes, with some internal logic on the computer and stored algorithms on color matching, contrasts, style guides etc, the Intelligent wardrobe makes recommendations on what clothing works with what. Furthermore, if the owner is really fashion-conscious, the intelligent wardrobe can also go online to check the latest trends for given fashions and styles.

With this information the computer analyzes the clothing the owner is currently wearing, like a shirt or blouse, and then the computer will offer recommendations.

But not just as a printout or on a computer screen. The system also makes suggestions right onto the mirror screen of what garment would fit the current outfit. This is the shirt you should wear, or with that shirt, this tie is what I recommend today. In the current model the mirror shows the text on the mirror, but soon this technolgoy will evolve to project the image of the actual suggested item of clothing to give a virtual mirror image of what is the suggestion.

The intelligent wardrobe (or mirror). This should go a long way to eliminate worries about mis-matched clothes on first dates etc.

And once the projected image is based on the true wardrobe of the owner, rapid cycling through clothing options can be made, selecting ties, scarves, etc without a need to dress and undress. As it only a digital image, as the owner says, "No, not that tie, show me the next" as it requires no undressing and dressing, various clothing options can be tested in a matter of seconds.

Coming to a home near you very soon.
Already reality in Digital Korea today.

Chapter V
Portable TV

Broadcast and Beyond

Image courtesy *IT Korea Journal*

> *"The mobile will be the main device to enjoy radio and TV programs anytime anywhere."*
> **Yun-Joo Jung, CEO KBS Korea Broadcasting System**

V
Portable TV
Broadcast and Beyond

On May 16, 2005, South Korea became the first country to launch digital TV broadcasts direct to cellphones on a standard called DMB Digital Media Broadcasting. Before this, many countries had offered "streaming" services to cellphones, as well as various analog broadcasts to pocket TVs and some advanced cellphones with in-built conventional TV tuners. Integrating the Digital TV set-top box into the cellphone was a radical innovation and by the time this book went into print we had a year and a half of early experiences with this, the next form of media consumption.

South Korea has of course traditional cable TV, broadcast TV, satellite TV, video, DVD and PVR (e.g. TiVo/Sky+) and IPTV technologies. These are not radically different from those in other advanced TV markets such as Japan, USA, UK, etc. The advent of the next digital TV platform, built into the cellphone, is the radical innovation in South Korean TV, and this chapter will look at TV in your pocket. Or more precisely, digital TV in your pocket.

A A LEAP AHEAD

So-called video streaming has been possible for cellphones on advanced networks for five years. In most countries in 2006 and 2007 when discussing cellphone TV, it is some form of video streaming or downloading of video clips. Streaming video to cellphones has inherent bottlenecks as it is constrained by spectrum and with the cellular network, is subject to congestion.

To optimize the service for cellular networks, the network operators/carriers use various coding techniques and limit the total streaming technology to produce TV-like experiences. They are a far cry from digital TV on a big screen, however, and have a very glitchy and jittery feel when watching television on them.

Broadcast TV is not constrained by the number of users in the network. If a receiver (i.e. TV set) can receive the signal, there is no limit to how many people can view the broadcasted show at the same time. So depending on the broadcast technology, far sharper images can be transmitted.

Digital TV to cellphones

South Korea moved into true broadcast form of digital TV on cellphones on the S-DMB standard (Satellite-Digital Media Broadcast). In May 2005 this was the first case of true digital broadcast to handheld and other portable receivers such as TV sets in automobiles and laptop computers. The best analogy is that the set-top box of the cable or satellite TV service in its digital version has now been integrated into the handset. Or to put it in another way, your cable TV decoder is built into the cellphone, with full digital services of course.

South Korean TV-cellphones of course have high-resolution TV screens to allow the full benefit of viewing digital TV. Top-end models include PVR (Personal Video Recorder) capabilities like TiVo and Sky+ and thus allow pausing live TV and recording favorite programs etc.

The launch was a major source of pride for the whole South Korean IT and Communication industry, as Daeje Chin, Minister of Information and Communication, was quoted in IT Korea Journal, Nov-Dec 2005, saying:

> *Satellite DMB is a mobile multimedia broadcast service delivering high-quality video and audio to portable devices and in-vehicle devices. During the APEC Summit, we will be airing English-language content provided through Arirang TV in addition to the regular programs currently in service. News programs will also be beefed up for the benefit of APEC guests, and more foreign films are scheduled for this period as well. By providing APEC participants a sound introduction to Korean IT, we are confident that a favorable*

perception of our technologies in these countries will give a big push to overseas expansion efforts by Korea companies.
Source: Daeje Chin, Minister of Information and Communication, *IT Korea Journal* Nov-Dec 2005

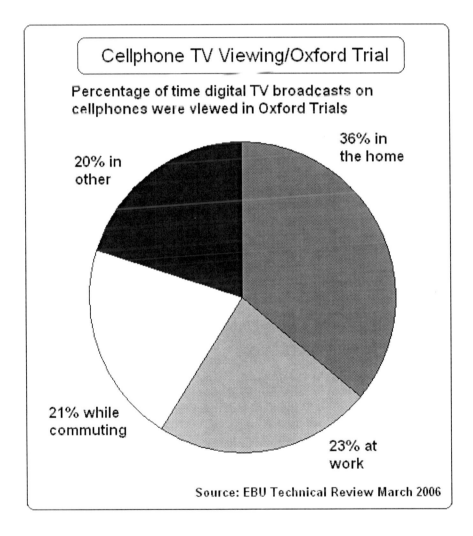

After the first network went on the air, by November 2005 a rival system on the T-DMB (Terrestrial Digital Media Broadcast) standard went into commercial production. By the end of 2006 there were over 2 million subscribers watching digital TV on their cellphones, or over 5% of the total South Korean cellphone user base had migrated to these, top-end phones and the premium services. It is worth noting that in South Korea there are no handset subsidies, so the South Korean cellphone users paid full street price of between 500 and 800 dollars to upgrade their cellphones to these newest top-end digital TV phones. By December 2006 the total count of DMB viewers on the two standards had reached 3.8 million.

Rich market for rich TV content

Korea is a very advanced TV market, which has been able to capitalize on Korean high broadband penetration and the Korean leadership in videogaming, the cartoon illustration industry etc. South Korean dramas and soap operas are also exported to many markets in Asia. Innovative companies like OnTimeTek with long history in mobile TV for 3G broadcast are building on IPTV and on the digital convergence of broadcast, internet and mobile telecoms with innovative solutions and concepts .

Since MiTV first launched IPTV in Malaysia, the convergence areas of TV and broadband internet are being developed in dozens of countries. In addition, digital TV is now evolving fast, with the UK in a leading position. The most compelling offering, however, comes from the proposition of consuming TV at all times in all places. The only way to do that is on mobile digital broadcast technology and ideally onto a cellphone.

Not your average pocket TV

We've had portable pocket-TVs since the early 1980s. They never took off as a major mass-market solution. Most pocket TVs were home, in some drawer, forgotten, when the sudden need might arise, like being stuck in a traffic jam or the train being late. But the cellphone is carried within arm's reach by all economically viable people on the planet, 24 hours per day. A 2006 Nokia global survey revealed that 72% of the cellphone owners use the phone as their alarm clock, so it is literally within arm's reach even as we sleep. No other device has achieved this level of global penetration. At 2.7 billion phones in use around the world at the end of 2006, there are three times as many cellphones as personal computers in the world, and twice as many people have cellphones as have a fixed landline phone, a credit card or a TV set.

That is why all other media are now eagerly looking towards the cellphone to emerge as a new platform and media channel. 16% of global music revenues is generated on sales to cellphones (mostly as ringtones and ringback tones) and 14% of videogaming revenues are already derived from sales to cellphones. We will discuss music and gaming in their own chapters later. However, this highlights the growing interest in the cellphone. Companies from the BBC to Google to Apple with its iPhone launch, have said that their future will be on the cellphone.

Not just snacking

We remind the reader of what the CEO of the largest Korean broadcaster KBS (Korea Broadcasting System), Yun-Joo Jung said, "The mobile will be the main device to enjoy radio and TV programs anytime anywhere." Many in the West may feel doubtful of this forecast, thinking that perhaps for short clips and "snacking" the phone is viable, but certainly not for consuming long content like TV series and movies.

Average amount of daily consumption of DMB digital TV on cellphones in South Korea is 129 minutes per day.

Source: MIC January 2007

For this the South Korean experience is clearly dispelling that myth, and more recent trials of digital broadcast TV to mobiles from Helsinki Finland to Oxford England confirm that based on the trials about 25%-30% of the trialists were fully happy to consume TV content on the advanced cellphones, for 90 minutes at a time. That means plenty of time to consume a full-length movie or a game such as football. While it sounds counter-intuitive, the survey data clearly confirms what is already seen in Korea - an increasing part of the population will watch even movies on the phone. What is critical is to see how clear the image is on digital broadcast to cellphones as it is on the DMB standards.

B THE SIX LEGACY MASS MEDIA

The first five mass media are all very mature, over 50 years old each. They are rather well known and understood. This book is not intended to be an analysis of the various other older media, and we will only list them here briefly to help put the sixth and seventh in context.

First of the mass media: print

The first mass media is the printing press. At about 500 years old, it gave us first books, then pamphlets, then newspapers, and later magazines etc. Print introduced the concept of owning the content (books) and advertising (newspapers) and subscriptions (magazines).

Seven Mass Media

First Mass Media Channel - Print (from 1500s)

Second Mass Media Channel - Recordings (1900s)

Third Mass Media Channel - Cinema (1910s)

Fourth Mass Media Channel - Radio (1920s)

Fifth Mass Media Channel - TV (1950s)

Sixth Mass Media Channel - Internet (1990s)

Seventh Mass Media Channel - Mobile (2000s)

Second media channel: recordings

The second mass media appeared about 1900, as recordings. The first recordings were "clay" records, eventually shifting that media to vinyl, and then digital formats expanding to videogames and movies on video tape and now DVD, which also increasingly offers TV shows as recordings. Like print, recordings are also a "buy-to-own" media. Recordings introduced the concept of the performing artist, or celebrity. Edith Piaff, Frank Sinatra, Elvis etc built their following through recordings. Recordings allowed political speeches, comedians and other shorter vocal items to be stored and sold.

> **Pearl - Interactive TV advertisements.** Tu Media has introduced interactive advertising which allows TV viewers to point to items on screen and then get more information on them. Like the shirt on the actor? Point to the screen, it pauses the TV show, and takes you to the shopping page.

Third media channel: cinema

The third mass media was cinema, from about 1910. This was the first "pay-per-view" format, and the first "multimedia" format even though sound did not come until years later. The cinema introduced continuing storyline films i.e. the cliffhangers (precursor to today's soap operas). Cinema also produced the world's first global celebrities, Charlie Chaplin etc. Cinema content was consumed in large groups (i.e. not privately). The advertising in cinema was shown before the main feature started.

Fourth media channel: radio

The fourth mass media appeared also very close to that time, essentially around 1920: Radio. This was the first broadcast media, where the consumption was a "streaming" concept. You did not own the content and the listener could not replay it (until technology emerged to capture broadcasts onto tape. For mass markets, this did not happen until the 1970s with Philips releasing the c-cassette, a full 50 years after the introduction of radio).

Payment on radio was either through a national radio license from the listening public as in much of Europe or by advertising as in America. Radio ran regular drama and comedy shows including continuing stories. Families would gather around the favorite broadcasts and listen together. Radio started to dominate other media - a pop music artist who was favored by a radio DJ would then become a hit on selling records. Thus for the music industry very specifically there became a close relationship between radio airplay and record sales.

Fifth of the mass media: TV the current giant

The fifth mass media is the biggest and most dominant to our culture today: TV. Introduced to the mass market in about 1950, TV did not really introduce anything new. We had multimedia in the cinema, and broadcast in radio. We already had the license and advertising based payment model also with radio. But bringing the visual and multimedia experience from the cinema to the

---

19% of Americans use cellphone based internet
Source comScore October 2006

---

home like radio, TV was riding on the most successful elements, combining them.

TV soon took over totally the news from cinema. It took over much of the drama series from radio, and the live sports broadcasts. Like radio, TV was first only a streaming proposition - if you did not see the episode, you missed it forever. Late 1970s Philips introduced home videotaping with its ill-fated VCR 1700 system and soon thereafter Sony's Betamax and the VHS consortium made video recording mainstream. Today recording DVD machines, home Hard Disk Drives and various PVRs (Personal Video Recorders) like TiVo and Sky+ are changing all that even further.

TV discovered the power of the celebrity, and soon shows emerged that promoted celebrity (e.g. talk shows) and those that propelled normal people into temporary celebrity status (e.g. game shows, reality TV).

Continuing storyline soap operas emerged killing the serial movie concept from cinema, and removing most continuing storyline drama from radio. After the advent of MTV Music videos, suddenly the radio and music recording connection was severed, and MTV became the deciding factor to a music artist's success.

Internet the sixth, is the first interactive media channel

So enter the sixth mass media, the internet, in the 1990s. The internet is a personal media, but not very portable. Yes, we can have the web on a laptop but typically most laptop web use is restricted to hotspots; most internet surfing is done seated in a fixed place. This is a very young media. Its main innovations were interactivity and search, which form the competitive advantage of most media content when ported to the web; you cannot search a pile of Wall Street Journals in your bookcase, but of course there is a Search field on the WSJ web site.

As a mass media, the internet could deliver similar experiences to most of all of the other five previous media - we can read books, magazines and newspapers online; we can view movies; we can listen to radio; we can

> **43% of South Koreans use cellphone based internet**
> Source NIDA September 2005

digital equivalents of recordings e.g. MP3 files, MPEG movies, TV content in clips and streaming; and yes, we can download the computer software, videogames etc. In addition, the internet has in its short life already very dramatically moved into each of those other established media, and often with cannibalizing and even arguably illegal ways (e.g. Napster and music, or Google currently on some copyrighted books).

The internet is based on philosophies of freedom and shareware and collaboration. Much of the content and applications are free or shareware etc. There are subscription models and advertising revenues also. In terms of content "ownership" it is a total hodgepodge, some stuff you can own, others you shouldn't and still others are very difficult to capture to own. There are many new media content formats from the MMOG massively multiplayer online game - virtual environments with literally millions of users co-creating the entertainment experience, to social networking such as blogging, wikis, chat rooms and online dating. A particular South Korean contribution is Citizen Journalism, which we will discuss in the Ohmy News Case Study in the Government chapter.

The other truly dramatic relevance of the internet is its cost. Almost nothing. You can become an internet service provider at trivial costs compared to any of the previous mass media; and to become a website or blogsite, you do not really need more than your connection. If you have free access say through the local library, you can become a web publisher for totally free. Moreover, in most countries the internet is very lightly regulated or not at all.

C CELLPHONE IS SEVENTH MASS MEDIA CHANNEL

So how of the 7th mass media? The cellphone was realistically only a voice device for the masses through the 1990s and only emerged as a mass media outlet with WAP, i-Mode and premium SMS from about 2000. The youngest of the seven mass media, it is by far the least understood. Like the internet, the cellphone is also an interactive mass media with search capability. Like the internet a decade earlier, the cellphone can do everything all of the previous media can do, making it a very serious cannibalization threat not only for the first five media but also for the sixth, the internet itself. This will accelerate with high-speed cellphones. South Korea is one of three countries where the majority of internet access is already from cellphones, with Japan and China the other two.

Has abilities beyond the first six mass media

The cellphone introduces new elements not possible on any of the previous six media including the internet, making the cellphone the most powerful mass media. First of all, the cellphone introduces a built-in payment mechanism. In fact users on cellphones assume all content is paid, whereas on the internet the assumption is the opposite: that content is free. Thus all payment models, subscription, pay-per-view and advertising are widely used. Also differing from the fixed internet, the cellphone is always carried upon the person, i.e. mobile content is always with us. The cellphone is very personal as a device, and its sharing is typically no more than between two people if listening to a song e.g. And at 2.7 billion users worldwide, the cellphone is by far the most widely spread mass media of them all.

Yes, there are more radios than cellphones, but those radios are in North America and Western Europe, built into our cars etc. In Asia, Africa and Latin America many more cellphones exist than radios. By the end of 2006 there was a cellphone subscription for 40% of the total population on the planet. Essentially every economically viable person on the planet carries a cellphone. Every one of them can do basic texting, basic mobile commerce, receive basic news, etc.

User-generated content

What is even more important to understand, is that the cellphone is not only the most powerful media consumption device, it is also the most powerful participation and media creation device. From bloggers updating their weblogs from cellphones to amateur paparazzi snapping pictures of David Beckham to the video clip secretly shot at Saddam Hussein's hanging, the cellphone allows media creation and participation. By 2006 one third of Americans who uploaded images to picture-sharing website Flickr, did so direct from a cameraphone.

> **SMS-to-TV related content in Europe alone was worth 900 million dollars in 2005**
> Source McKinsey October 2005

As the cameraphone penetration increases, as it has reached over 99% in South Korea, and as the cellular network gets fast enough for rapid connectivity, and most importantly, as long as the carriers/mobile operators don't charge an arm and a leg for picture uploading, then almost all pictures will be sent directly from cameraphones to picture sharing sites. Again we see that already in South Korea, where 90% of all pictures transmitted from cellphones end up on Cyworld.

Cannibalizing threat to older media

The cannibalizing effect of mobile as the 7th mass media to the previous six media seem very strong, with magazines and newspapers recruiting content - pictures, and SMS text messages from readers; radio inviting text messages from listeners; TV reality shows using SMS voting; and even internet content being paid for by premium SMS such as the ten million users know who use the youth online playground Habbo Hotel.

A special service provider category is appearing in wireless telecoms in most markets, called the MVNO, the Mobile Virtual Network Operator with Virgin Mobile perhaps the best-known example in the UK, Australia, USA and Canada. The MVNO is a specialized wireless carrier, which does not own or operate a cellular network but rather leases network capacity on a wholesale level and then resells it under its own brand. In the USA recent examples of MVNOs include 7-Eleven, Amp'd, and Disney. ESPN was an MVNO but pulled out of the business, and Helio is one that has a strong South Korean connection, bringing many South Korean services such as Cyworld to America. Through the advent of MVNOs suddenly the cost of entering the wireless telecoms industry has fallen dramatically.

All seven will continue

We should mention that even though there are six "new" rivals, the oldest mass media - print at 500 years - is still very healthy, and none of the seven is seriously at risk of ending as a commercial opportunity. Yes, some pundits have suggested that the paid daily newspapers will disappear sometime during this century, but even those do not suggest all print media will disappear. We still buy hundreds of magazine titles, read free newspapers, and more new book titles were published in 2006 than at any time before. Even the oldest mass media, print, has a lot of life in it.

So while the internet and the cellphone show very powerful strengths to cannibalize areas of the established five media, all seven mass media will co-exist for a long time to come.

D IS DMB PART OF SEVENTH MASS MEDIA?

DMB is an interesting opportunity. It is a digital broadcast technology so it is certainly a close cousin of traditional broadcast TV. However, not unlike IPTV services on the internet, which are starting to diverge from the basic formats of traditional broadcast TV, so too will DMB and other digital broadcast technologies to cellphones allow innovation, for TV to move past the box at home. DMB alone does not inherently allow interactivity, but if TV services are integrated with the interactivity of the cellphone, then true innovation can be unleashed.

Interactivity is key

Here South Korean broadcasters and carriers have innovated and today DMB has numerous types of interactivity from overlaying content to the TV image, the use of widgets, menus simultaneously on top of the broadcast content and many exciting new forms of opt in interactivity content provider/advertising

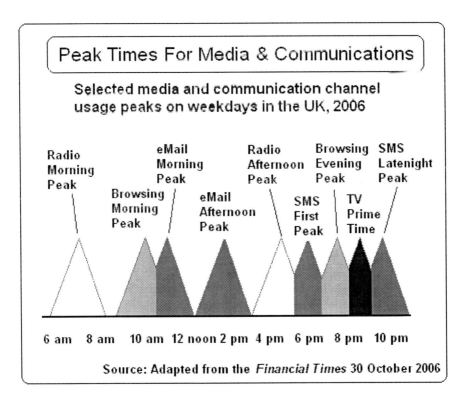

Independent of the broadcast TV standard for mobile, Korean handset makers continue to innovate in this space as well. On the rival digital mobile broadcast technology DVB-H (Digital Video Broadcast Handheld) which is popular in many European markets, South Korean providers LG and Samsung have been introducing advanced and very popular handsets for example in Germany and Italy.

So is DMB fifth or seventh mass media? We would put it like this. DMB alone is not enough. Just TV programming on DMB will be similar to other digital broadcasts and fall under the fifth mass media category. However, if integrated to capitalize on the capacities of the cellphone, then DMB becomes part of the seventh mass media.

It also should be noted that the seventh mass media is not limited to TV content. We can have print content, web content, music content, gaming content etc on the cellphone as well. Therefore, the seventh mass media is as much a multimedia content platform as the internet is, perhaps even more, with its built-in payment mechanism. We have many examples in this book about content formats for the seventh mass media beyond TV on cellphones, such as Ohmy News the citizen journalism newspaper, Melon the music service, and Cyworld the social networking site just to name three. Each of them is case studies in this book as well.

> **By 2005 out of Japanese internet subscribers 66% had migrated to broadband**
> Source ITU 2006

E   OTHER TV AND VIDEO CONCEPTS

There are many other video concepts for the cellphone as well as other evolution elements to TV. We will touch upon these briefly here as well.

Non-broadcast video content to cellphones

Mobile TV is only five years old, as two innovations were launched simultaneously in 2001. In Finland SMS-to-TV chat went live in 2001, while MTV launched its radical *Videoclash* in the UK - the program where viewers could decide what videos to see next, and vote via cellphones. Videoclash also premiered in 2001. Then for the next three years most analysts, forecasters and pundits in the cellular telecoms and digital TV areas suggested that the future of TV on cellphones would be "snacking", i.e. consuming short clips, such as football highlights, CNN news clips, movie previews and music videos onto cellphones. Certainly these will form part of the cellphone video ecosystem, but by 2006 it was obvious these would not be the bulk of what is consumed.

Mobisodes

By 2005 the young cellphone-TV converged industry started to see true innovation, in the area of mobisodes for example. Mobisodes are TV programs shot in short form and considering the cellphone experience, i.e. with lots of close-ups, not large crowd scenes, etc., and usually are shorter in length than 30 minute and 60 minute broadcast TV formats. Mobisodes are usually sold per episode or per subscription to a season of the show. The first mobisodes were introduced in Italy, Netherlands and Japan.

In Spain Endemol, the creator of the *Big Brother* format, developed for Telefonica a comic strip series of soap opera stories where viewers could vote on what happened next. Every three episodes the viewers were asked to

> **By 2005 out of South Korean internet subscribers 100% had migrated to broadband**
> Source ITU 2006

suggest where the plot should go, who should die, etc. This is difficult to do with real actors but very easy to do in animation. And since the cellphone includes the built-in interactivity channel - and can generate revenues from SMS text messaging votes - this kind of innovation becomes possible.

Meanwhile MTV in Europe introduced its *Head and Body* series of quirky and often very edgy humor to mobisodes. A series considered too rude for cable/satellite TV broadcasts of MTV, the cellphone TV allowed a more targeted focus to the show. Some mobisode concepts failed, such as the mobisodes around the internationally popular show, *24* with Kiefer Sutherland. As the mobisodes did not feature the main stars of the TV show, fans soon rejected the mobisode variant.

Simulcasts

Another area where TV/video content on cellphones can thrive is simulcasts and multicasts via 3G phones. Robbie Williams, a very popular British singer, promoted the launch of his new CD with a premiere concert in Berlin. He attracted over a million viewers on 3G cellphones in various European countries joining to watch the simulcast of the live concert. At the MTV

Europe Awards the mobile MTV channel went back stage and shot exclusive footage that was only seen on cellphones. This was intended to be watched in parallel with the broadcast show live on the "normal" TV screen at home, and then the "spycam" onto the back stage via the 3G phone. At *Big Brother* houses around Europe it is now commonplace to have exclusive spycams to see what is happening in other rooms, while the TV broadcast show will typically show the room with the most housemates. These are the kinds of new services that become possible when traditional audience concepts of viewership can be abandoned and viewers are actually charged per view. The cellphone is the optimal channel for this kind of TV consumption as it is always present within arm's reach when we watch TV.

Perhaps the most advanced area of cellphone TV capitalizing on the true strengths of the seventh mass media is user-generated TV content. There are many chat boards and interactive TV games that use SMS text messaging - and recently also MMS picture messaging content from viewers. However, in South Korea the broadcaster Tu Media has introduced its concept of user-generated broadcast TV. Viewers are invited to send TV clips to the show. We will discuss this in more detail at the Case Study at the end of this chapter.

What of Video-On-Demand (VOD)

Another promising development on the side of digital TV broadcasting and cable TV is Video-on-Demand (VOD). Many analysts and pundits suggest it is the killer application that will put video rental stores out of business, and drive the usage of broadband and digital TV adoption. Launched in many countries, South Korea saw its first VOD offering back in 2001 on cellular networks with SK Telecom.

Pearl - Soap opera previews, on 3G cellphones. Really addicted to your soap opera? Need to know what will happen today. In South Korea every morning the networks offer previews of what will be on the soaps that day. Premium video content on 3G cellphones of course.

On the broadcast TV side, Korea Telecom introduced the more common TV variant of VOD in 2004. Today VOD services are offered by several companies and they form the marketing focus of the triple play provider Hana TV by Hanaro Telecom, which bought Celrun TV. The Hana TV package includes internet, telephony and TV with VOD and other multimedia services.

The VOD concepts are also seeking a market opportunity with video podcasts, various home entertainment solutions and video innovations such as Slingbox. In the near future there is likely to be a lot more redefinition of these concepts around the commercial viability of personalized video streams for VOD consumption. Broadcast TV such as digital terrestrial, satellite, and cable television use a single channel to deliver a given selection of content, say a movie or TV series to all viewers (or subscribers) in the broadcast or cable reach of that TV broadcaster.

As long as VOD solutions require individual cable/satellite/digital streams to be allocated per subscriber for one viewing, they tend to be inefficient in distribution and require prices that many are not willing to pay. However, this area is likely to evolve a lot with the technical innovations that are now being deployed.

Hana TV

In June 2006 Hana TV launched its triple play service of internet, telephony and TV/multimedia. The advanced TV and multimedia offerings are the spearhead for Hana TV including Celrun TV an established VOD provider in South Korea. In addition to traditional internet, telephony and cable TV (and VOD) services, Hana TV also features karaoke and gaming, both very important for the South Korean market. The audio and video quality on Hana TV matches the demanding South Korean market with HDTV quality video. Among the VOD offering, Hana TV rotates 30,000 titles of content.

Turning off the TVs

In just about all areas of TV evolution South Korea leads. From digital broadcasts to cellphones, onto IPTV and VOD. Perhaps the most exciting areas come from convergence and discovering how the 7th Mass Media is different from the previous six. For that stay tuned to South Korea

Case Study 4
Tu Media

Tu Media was the first DMB broadcaster on the S-DMB standard. With the SK group as partial owner, Tu Media has benefitted from the start out of the full cellular telecoms side of interactivity via SK Telecom. Tu Media has then been able to explore very extensively the innovative sides of the DMB technology and true concepts into the 7th Mass Media

Tu Media is a subscription service which recently lowered its charge from 13 dollars to 11 dollars per month. There is also an annual subscription option which costs about 10 dollars per month. With a million paying subscribers in about two years, Tu Media is rapidly approaching its break-even point as a business.

World Baseball Classic tournament in 2006 where the South Korean team beat heavily favored teams from the USA and Japan. As the games were mostly broadcast during the day, this doubled sales of handsets and subscriptions.

Rolling User-generated Videoclips

A good example of a Multi-user mobile TV service is viewer participation in the form of SMS-to-TV programming. Launched in Finland in 2001 as a chat board broadcast nightly on TV, where viewers would send premium cost SMS text messages and thus see the chat board on live TV, the concept has since evolved to include SMS-to-TV dating as in Italy, SMS-to-TV games as in Malaysia and SMS-to-TV Rap the latest hit in Finland. But the most advanced concept of premium user-generated content on the cellphone, broadcast live on TV, comes from South Korea of course, on the Tu Media network. In late 2005 Tu Media introduced rolling user-generated videoclips. Any viewer could use their cellphones to capture a video clip, and then send the videoclips as premium-cost video messages to Tu Media, and moments later these would be broadcast live.

Its a little bit like YouTube in collecting user-generated video but now those videos are also shown on the broadcast channel. So imagine your child having a 6-year birthday with plenty of other 6-year olds singing and dancing and eating cake. What better way to thrill the 6 year old, than shoot a bit of video (which no doubt your child has seen you do plenty of times) but then send the clip to the TV station and turn on the TV. In a few minutes your child is on the air, blowing out the candles of the cake.

Of course to respect local laws on decency and broadcast content, the clips are screened with human editors viewing all clips and removing any objectionable content obviously, such as nudity and profanity.

In some way this is the optimal merger of the YouTube concept of user-generated video and broadcast TV. As the broadcaster can charge for sending the clips, they make money on the proposition. Users love it as they can be on "real TV" rather than the amateur DVD that the techie-uncle made, or "just" on a video sharing site on the net. This kind of innovation becomes possible with the the 7th Mass Media.

Chapter VI
Online Shopping

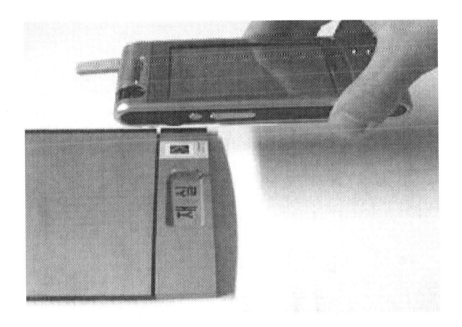

Money Turning Digital

Image courtesy *IT Korea Journal*

> *"I didn't lose my mind, I sold it on eBay."*
> **Anonymous**

VI
Online Shopping
Money Turning Digital

What happens to money in a fully digital world? The near future of commerce and the very nature of money in a digital era can be seen clearly in South Korea today. The biggest bank in Korea says their competition is not American Express and Visa and MasterCard, or the other traditional banks. They say their competition comes from money on cellphones. Globally nearly twice as many people own cellphones than own credit cards. If it takes an age limit of 18 years to get the first credit card, many teenagers get their first consumer credit when they get their first post-pay/contract cellphone account. For many Europeans now the first phone is received at age 8. That is ten years before they are eligible to a credit card. In Korea there are five separate cellphone payment mechanisms and already over half of Koreans pay regularly using their cellphones. We do not mean buying the odd ringing tone or phone voting with a reality TV show. We mean to pay for regular groceries in the supermarket, to pay for petrol at the gasoline station, to pay for subway train fares etc. The value of payments in South Korea that were paid by cellphones in 2005 exceeded 1 billion dollars.

A MOBILE WALLET

Therefore, it starts with the mobile wallet, the concept of your cellphone incorporating credit card and bank charge card abilities, as well as other functionalities of the wallet such as pictures and ID cards.

Following the lead of the Philippines, South Korea was among the earliest countries with mobile payment systems on cellphones with the first services going live in 2003. All three carriers have full mobile banking and cellphone credit card systems. Over 2.7 million South Koreans use mobile banking via wireless carrier services on SK Telecom' Moneta, KTF's K-Bank, and LG Telecom's Bank On. The growth in users is 50% annually. Five separate credit card systems operate on the three cellular networks. SK Telecom's "Moneta" service has over 1 million subscribers, KTF's "K-merce" service has over 500,000. There are more than 470,000 locations nationwide that will accept m-payments from m-banking and mobile credit cards on cellphones.

Moneta

SK Telecom the biggest wireless carrier/mobile operator has its in-house mobile payment service branded Moneta. Moneta uses a smart chip on the cellphone, which is linked to a credit card account. The phone becomes a contactless style "card" at the point of sale. The phone transmits the necessary info via radio frequency or infrared technology. Thus no cellular network traffic is generated, which differentiates the typical modern mobile payment systems like those in South Korea and Japan, from earlier ones that depended on every transaction generating an SMS text message, like in use in the Philippines, Nigeria and South Africa. The Moneta system can be programmed with password-free transactions to a value of low payments such as paying for public transportation etc where the user will not need to give a separate pin code authorization. But for larger payments like paying for electronic goods or a hotel bill, a pin code is asked.

No SIM card

European and many Asian readers will perhaps think that the smart card functionality is built into the SIM card as on GSM phones. However, in South Korea most phones operate on CDMA standard networks where SIM cards are not used. The operators advised the handset manufacturers to insert a smart card element to the CDMA phones for South Korea so they could

Chapter 6 - Online Shopping

then build smart services around these. This is how South Korean CDMA phones have these smart card slots.

On the newer WCDMA/HSDPA handsets that conform with the "3GSM" evolution networks that now are also spreading in South Korea, this problem does not exist, as they use the SIM cards familiar on GSM networks. Information from these smart chips for mobile banking and commerce can be transmitted in either of two ways, either on short-range radio or via the infrared port on the phone. A particular benefit of providing the smart card to the phone user is that when the person applies for credit, the credit card utility can be enabled over the air within seconds after the credit has been approved. With traditional credit cards, a mailed card would typically arrive only about a week after credit has been approved.

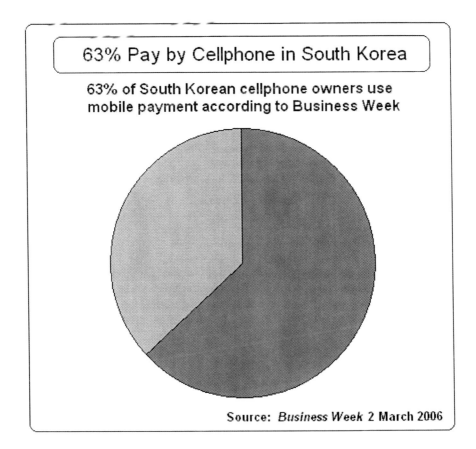

Do you want plastic with your credit?

A particular quirk of digital money is that the plastic becomes redundant. South Korean credit card companies now ask their new users if these require the no-cost plastic card to be also mailed to their home address. These are needed essentially only when the Korean leaves the country and encounters "old fashioned" credit card readers who need to swipe a plastic card. So for purely domestic credit card use, plastic is no longer necessary in South Korea. Makes you think? How soon will this happen in the rest of the world, that credit card functionality remains but the "plastic" vanishes.

25% of all Visa Cards in South Korea are provided via cellphone
Source: Mobile Payments World Q3 2006

Again typical of the organized nature of how South Korean society embraced the opportunities in digital money, the Korea Mobile Payment Services (KMPS) was founded back in 2000. With 250,000 merchant customers it is the umbrella organization to handle payments, provide support to the mobile banking and mobile commerce initiatives, and act as a spokesbody for the industry.

B OTHER ELECTRONIC PAYMENTS

The cellphone is not the only way to use digital money in South Korea today. Some of the obvious other ways are for common methods as online access to traditional banks, using credit cards online and setting up online payment accounts such as Paypal. Digital Korea of course offers even more options.

Postal Banking

Cellphones are not the only digital money initiative in South Korea. The post office for example has a large presence in the banking sector. The Post offices were recently upgraded to an ATM digital backbone that allows secure money transactions between the offices. 25% of all bank accounts in South Korea are managed by the Post offices.

Cyworld is also a virtual shopping world

We discussed Cyworld in previous chapters to a great degree, and also discuss it in the music chapter later. Therefore, Cyworld as a virtual playground, blogging and social site should be very familiar to our readers by now. We want to note that Cyworld has become a vast commercial world, in addition to the social networking site. In many ways the extent of commerce inside Cyworld is a precursor for many of the better-known western massively multiplayer games like Worlds of Warcraft and Counterstrike, and virtual reality environments like Second Life and Habbo Hotel. Cyworld today has more than 30,000 separate business entities with a virtual presence, and they have generated over 500,000 items of digital content, which are for sale inside Cyworld. The content for sale includes virtual properties for consumption inside Cyworld such as Levi's blue jeans and Nike sneakers for your avatar as well as virtual sales of goods or services consumed in the real world, such as selling airline tickets or insurance online. As we will see more in the music chapter, Cyworld has become the largest outlet for music sales in South Korea.

What level of money transactions is involved? Cyworld economics are growing at a dramatic rate, so whatever we write is going to be out of date by the time this book is published, but for context, in the summer of 2006, the daily value of Cyworld transactions was worth over 450,000 dollars. This is about 15 million dollars worth of business per month, or 180 million dollars at an annual level. This from a virtual e-commerce market place that was not set up from the start to be an e-commerce site like eBay or Amazon.

Convergence with the finance industry

After the credit card industry noticed that more people have cellphones than credit cards, and that cellphones are being used by kids too young to qualify for a credit card, they became very eager to "jump into bed" with the carriers. SK Telecom in Korea was the first carrier to issue its cellphones with Visa credit card functionality in 2002.

Today numerous wireless carriers around the world do that from Telenor in Norway with Visa to Smart in the Philippines with MasterCard, while NTT DoCoMo in Japan has started to offer credit card functionality as an add-on benefit to its Felica mobile payment system. In some developing countries telecoms carriers are more trusted than banks, so in places with high amounts of corruption and regular banking scandals and bankruptcies, such as Nigeria, the cellphones handle more banking transactions than real banks.

In addition, in many countries where credit card activities have a lot of fraud, banks and credit card companies offer real time alerts via SMS to inform users of credit card use. These kinds of services are very popular in South Africa for example. In addition, small person-to-person cash loans are now handled via mobile networks like offered by Telefonica in Spain. Clearly the banking and credit card industries are experimenting with the "mobile wallet." Still for all their innovativeness, the most advanced mobile wallet country continues to be South Korea.

How does it work

Card Authorization Terminals (CAT) are increasingly wireless in South Korea such as those by Nuri Telecom, which uses the ZigBee (802.15) short range standard for radio communications and includes a digital signature function. The CAT can be used to authorize and settle credit cards, debit cards and check cards as well as printing receipts. It is used for example at gasoline stations, department stores and restaurants in South Korea.

Automated digital interactivity

SK Telecom is developing an RFID (Radio Frequency IDentification) based extension to its mobile commerce application on the cellular telecoms network. This is how it could work: While walking in a town in South Korea, you notice a billboard with an advertisement, say one for flower delivery. The billboard is embedded with the RFID radio chip, which is seeking nearby contacts.

As your cellphone comes within reach, you can connect to the billboard and it provides the current offer from the florist nearby. A few clicks later you have ordered your flowers, which are billed to the account, associated with your phone. The florist ships the flowers as requested. Simple. The point here is that SK Telecom does not earn money on the actual RFID traffic between your phone and the billboard, but gains extra mobile commerce traffic when your phone dials up the florist and places the order.

C ONLINE AUCTIONS

Western readers are familiar with eBay as the prototypical online auction site. Auction sites such as Auction.co.kr. are yet another example of how Korean citizens think in "bballi bballi" (hurry hurry). Rather than go back and forth several times haggling to a given point, most items on sale on South Korean auction sites simply display the final price. Take it or leave it. This typifies the knack for South Koreans to cut to the chase and pursue efficiency and next generation behavior.

Skip the haggling

Most products sold on the Korean auction sites are sold by traders so there is little need for auctioning. There is one price. Sounds like a paradox yet this is emerging as the pattern on many repeat purchase items on American and UK auction sites as well after a given price level has been achieved. The Koreans tend to be very savvy price-conscious buyers so they will rapidly notice what is a good bargain and what is not. Equally the sellers can detect what price moved inventory and what does not, so they too can optimize their transactions through this mechanism. While these kinds of retailers do technically exist on auction sites, most transactions actually bypass the whole auction philosophy and its lengthy bargaining process.

> **Value of South Korean m-commerce transactions for non-telecoms goods will exceed $1 billion is 2006**
> Source: *Business Week* 2 March 2006

As most transactions are just straight sales and not auctions, consequently the market is much more efficient in terms of delivery. More volumes from same sellers means that regardless of price and almost regardless of the size and weight of the item being bought, most products will be delivered for a charge of about 2-3 dollars, the next day. Compare that to Western auction and e-commerce sites where delivery prices often make up the margin on the product and are quite costly and very fragmented.

In addition, the perceived risk for consumers is lower as buyers prefer to buy from real commercial dealers rather than individuals. Equally businesses do not have to pretend it is a private sale when it is not.

Always digital

Payment options in auction are by direct payment transfer or credit card, as cheques do not exist in Korea. This is transparent to consumer as it is a nationwide electronic transfer system –therefore the system has high trust by its users. The payment mechanisms on the web purchases in South Korea have high security and low fraud as every login forces an update to the software certificate.

In South Korea the web payment business model is based on sellers paying the commission or transaction costs to the bank/credit card company and therefore the transaction costs are included in the price.

14% of British cellphone owners buy MP3 music
Source: Telephia January 2007

Designed in Ubiquity wins again

In Korea two options for delivery are used, either pay the delivery person directly or pay delivery in advance. Many sites offer free delivery, as in South Korea the expected and demanded level of customer service is very high. Similarly despite the efficiencies of online selling, if there are any enquires of questions on the order, the seller will call up the buyer and provide assistance via the phone.

D COMMUNITIES DOMINATE

There are very popular "Buy it together " sections on websites where if volume of customers is high the price is lowered. To facilitate this, the customer count is visible on the interactive website. There are special prices for buying it today and some sellers offer temporary discounts to move inventory. Auction companies also have online events so customers can for example register for events where discount vouchers are distributed to attendees. While such online events are infrequent and on an irregular schedule, they are quite popular and drive traffic.

No second hand

It should be noted that South Korean commerce does not have a strong heritage of second hand goods and this is reflected online. South Koreans prefer to buy new.

Recently credit card debt has climbed in South Korea due in significant part to government intervention where tax incentives were provided to boost credit card transactions. This counterintuitive policy by the government was in reaction to the black market economy that was cash based so tax incentives were introduced to move cash transactions to credit cards and other electronic payment mechanisms.

> ## 45% of South Koreans buy MP3 files to phones
> Source NIDA September 2005

Demanding customers

South Korean consumers are notoriously demanding and vocal in voicing criticisms of products and services. This also translates to their negative comments online. South Korean online retailer services and reputations become even more sensitive to what consumers are saying about them online. In our experiences the South Koreans are significantly more likely to make critical comments than web users in other developed countries. When added with the intensely collaborative nature of community sites and blogs, the clustering of bad comments will rapidly influence any online retailer. If in the West on auction sites it is "Buyer beware", then clearly in South Korea it is "Seller Beware".

Digital footprint

The digital footprints of bad customer experiences linger very long in the blogosphere and on social networking sites. Even though business culture and local habits do matter, in the online world digital reputation is particularly important. The more there is true competition between providers and a larger market, such as that online in digital South Korean then the market has a long memory and no mercy. Digital citizens will soon see through inferior offerings and easily verify customer service reputations.

> **23 Million South Koreans, 46% of the total population make payments using one of the five cellphones mobile payment systems.**
> Source: Business Week 2 March 2006

This in some ways explains why the bar is set so high in South Korea for many high performance aspects of their ICT experience. Sadly in Europe and North America many initiatives do not succeed because users are sold inferior services on the basis that they may not know what good is or correspondingly services are not marketed at all based on a superficial reason such as that will not work here because we have a different culture. These gaps are closing fast as worldwide the tech savvy youth learn to navigate the digital universe and uncover bargains and verify capabilities of retailers just as easily as they discover hidden treasures in multiplayer games and navigate virtual worlds. Meanwhile in South Korea these newly enlightened consumers are clairvoyant master-shoppers perfectly versed in the real market value of what they pursue, knowing what it can do, who are its competitors and where it can be bought for the least.

That is what makes South Korea such an exciting but merciless environment for new marketers. The status of South Korea as a test bed status will grow as increasing amounts of service and technologies are joined into the digital Korea framework. South Korean customers become even more powerful through a more potent infrastructure reaching ever more deeply into the country and society.

Naver

As the consumer becomes more aware, there is an increasing value to good and up-to-date information online. One of the answers to this need is the shopping support site Naver.com in South Korea, which is an intelligent search engine, optimized for shopping. Naver has search using common language and phrases along with FAQ (Frequently Asked Question) sheets for common products and services.

The more advanced features depend on the knowledge based systems supporting Naver.com. Users are invited to respond to several series of questions to fine-tune the knowledge base for the consumer and adding to the total ability. The more users reply to questions, the more they earn points that can be exchanged for gifts or money. The system verifies and validates respondents so that given experts will receive incentives to give answers in their fields of experience. This is like Amazon book reviews, except that Naver.com removes the random ignorant reviewer in favor of those who are competent at reviewing that given type of book (or movie, album etc). Naver.com is also a major portal and its knowledge-based search is a strong tool to anchor loyal customers.

Another particular South Korean online phenomenon is the emergence of recommending societies, which recommend and reject products and services. The endorsement of a recommending society can be very important for a new product and obviously the rejection of a product by a recommendation society results in significant reduction in sales.

Silver digitals

What are older people doing and buying online. Both online and offline self-improvement is the key and there are mass market devices that cater to the old or are health related. Services and products range from electronic acupuncture to cellphones that feature diabetes checks etc. Many elderly South Koreans use the web to co-ordinate group traveling in particular for overseas trips. Similar to the previous generation of Japanese tourists, South Koreans tend to be seen by hoards in popular tourist destinations around the world.

A particular impact comes from Ohmy News the citizen journalism newspaper that we will discuss in more detail in the Government Chapter later in the book. Ohmy News has altered the media landscape in South Korea. In only four years, and with over 90% of its content generated by citizen journalists rather than professionally trained journalist staff, Ohmy News has become the fourth most trusted news source in South Korea.

With over 50,000 registered contributors, the newspaper has effectively a whole army of reporters where most TV stations, radio news and newspapers can only support a professional staff dozens of so reporters at best. Thus Ohmy News tends to report on new trends and sudden changes in South Korea much faster than its traditional media rivals do. Koreans are more likely to believe and be influenced by what other "average citizens" say especially on issues relating to shopping, so reports in Ohmy News can impact a given service or product's sale very rapidly. Opinion can be swayed fast and furiously and not necessarily via traditionally corroborated media sources.

Family matters

Some features of the South Korean society bear a major impact also to shopping. One is the traditional role of the husband and wife in South Korea. With very long working hours, the South Korean concept of the husband is that of the salary man. The husband works hard and then brings the salary to the wife. This means that the wives do the majority of all shopping in South Korea. The husbands tend to be at work until the shops have closed.

> **Pearl - RFID chip in wrapping paper?** Your florist wrapping paper includes RFID chip. You will be offered discounts to flowers, a link to the florist website and ability to send flowers, set up reminders, and pay for flowers, all via your cellphone reacting to the RFID in the wrapping paper.

With this, the wives are very astute in shopping and tend to control the family's disposable income as well as the bankbook, while handing the husband an allowance. When the internet started to penetrate deeply into South Korean society, the government noticed it had to train the women in how to shop online. This may seem counter-intuitive, women needing to be trained to shop, but because of the major role women play in shopping, for e-commerce to take off, the South Korean women did need to learn to shop online. Today this is no longer necessary, obviously. However, because of it, the South Korean user statistics may seem particularly skewed towards female usage of online shopping services when doing international comparisons.

In South Korea also the influence of Confucianism is very strong and a deep bond exists between parent and child. This may result or example in parents sacrificing when the teenager reaches working age and the parents take a diminished standard of living to allow the young adult to establish a good standard of living and start a career. This is then returned some years later, with the child returning the "loan" or support by diverting part of the income back to the parents.

South Korean parents are also obsessed with educating their offspring - which helps explain why South Korean student performances in mathematics, languages and sciences rank near the best in the world almost on par with Finland. Because of the focus on education parents are very willing for example to buy laptops to their children and high speed internet access etc.

South Korea has mandatory military service for all males, which lasts two years. This also means that boys tend to remain at home into their twenties before leaving the home of the parents.

Case Study 5
2D Barcode

The 2D barcode (square scribble, looks like a fingerprint, see above) is rapidly revolutionizing cellphone access to the internet. This in turn introduces countless new opportunities for rapid web access while on the go and often while doing other things simultaneously.

The traditional way to enter web addresses is by typing on a keyboard on a PC, or multi-tapping on a phone keypad. Web addresses tend to be lengthy such as

www.thisismywebsite.com

and even worse, specific campaigns may have longer variants, such as

www.thisismywebsite.com/campaigns2007/promotion

On a PC keyboard this is not overly cumbersome. Often those addresses are hidden underneath Hypetext links on web pages. But on a cellphone, it is most inconvenient to spot that type of web address in a magazine ad, and then start to tripple-tap.

The 2D Barcode is the almost magical answer to unnecessary typing of such data as web addresses. The 2D barcode embeds the data. A cameraphone with the appropriate software will scan the 2D barcode and momentarily display on the phone screen, the actual text embedded within it. Now any web address or other information such as personal data on a business card, etc, takes only a cameraphone scan and about one second of time. An enormous improvement to user convenience. 2D Barcodes were first introduced in South Korea in 2003 and launched next in Japan in 2005 to also great success. They are now starting to appear in the West with for example Nokia offering the 2D barcode reader on its top line phones.

The service applications for 2D barcodes are exploding in South Korea. Magazine and newspaper ads in South Korea now prefer to display the 2D barcode rather than the traditional web address. Posters at bus stops use 2D barcodes to provide links to web sites, video clips, coupons etc to promote access and provide further digital assistance. With our interviews for the book we met with the KW Park, the COO of Iconlab, which one of the pioneers of 2D Barcode technology in South Korea. Mr Park said:

"2D barcodes are particularly well suited for small business marketing and promotion use such as in advertising. It is cheaper to provide the 2D barcode which adds much more information to the consumer than what is possible in leaflets, brochures etc. The multimedia content of a videoclip, web page, sound file, specification sheet, etc can be all provided via an access through the 2D Barcode."
KW Park, COO of Iconlab

A good example is the 2D barcode on a wine bottle label. The consumer can go to the vineyard website, find the exact data on that bottle vintage, its reviews, etc. Yet the 2D barcode itself is not intrusive, such as adding a second full-size label with printed text of the same information.

The consumer response to 2D barcodes has been overwhelmingly positive. Customers love the speed and ease of access. As 2D barcodces then appeared in countless uses from airline tickets to clothing labels, South Korean consumers rapidly became exposed to them and had ever more daily reasons to use them. Judging by South Korean customer opinions, 2D Barcodes are likely to become one of the biggest innovations in consumer interactivity and web access.

Chapter VII
Electronic Government

Broadband Bureaucrats

Image courtesy *IT Korea Journal*

> *"Our main concept is every citizen can be a reporter. We put everything out there and people judge the truth for themselves."*
> **Oh Yeon-ho, Founder and Editor of Ohmy News**

VII
Electronic Government
Broadband bureaucrats

The internet has revolutionized almost every industry and human activity from consuming music and books to user-generated content and community behavior. Most industries from the media to advertising to banking to retail, have rapidly embraced digitalization for a competitive advantage. Governments overall have been rather slow at pursuing "e-Government" initiatives. South Korea took an early interest in digital benefits in government work and offers today a vast array of complex services for South Korean citizens. In various surveys the South Korean e-Government initiatives come out on top.

Whether you want to file your taxes, check on your student exam scores, search for missing relatives in North Korea or provide a comment on proposed parking changes near your home, all that can be done digitally and across various access devices.

A STARTS WITH THE CITIZEN

The South Korean Minister of Information and Communication Jun Hyong Roh puts the focus of government initiatives into very human terms explaining the intentions to create the digital Korea. Minister Roh clearly stated their position saying, *"An information society is a people first society and investment in human resources is important."*

All of South Korea's initiatives around digital Korea have kept the human focus. It starts with education. South Korea has the world's highest university attendance rates in the world - 80 per cent of high school graduates continue onto further education. With a work ethic valuing education and parents supporting studies, South Korean rankings in global comparisons of skills in mathematics, languages and sciences have steadily climbed toward the best in the world now rivaling those of Finland on the top.

Digital society

South Korea's government services start with the eGov portal. This impressive collection of most of the main government services combines over 500 separate services that are all integrated and executable online.

50% of all phones sold globally in 2005 were cameraphones

Source: SonyEricssson 2006

Services and departments from healthcare to education, from taxes to law enforcement, etc are all covered. All services offer the digital files of government publications, and digital access to the related databases. Typical of the high rate of broadband internet and highly advanced cameraphone adoption, nearly all of the government services already offer audio and video clips to support the services.

The South Korean sites are highly customizable by the citizen, with a majority allowing the user to manage his or her own activities. In addition, most sites offer cellphone or PDA access, and nearly all allow visitors to sign up for e-mail updates. Also notable are the interactive features available for users such as posting comments, participating in surveys etc. Virtually every site contains a prominent guest book or forum as well as the option to petition the particular department. Where most traditional governments in the industrialized world were accused of hiding behind bureaucracies of tens of thousands of civil servants per department, in South Korea each bureaucrat can be reached directly in their given area of responsibility.

A cunning plan

Typical of how methodically South Korea approaches its goals, the government has worked on tight schedules and ambitious plans, with broad support and ample resources to achieve its objectives. After successfully completing eleven major projects initiated in 2001 to form the basic foundation for the electronic government, in 2003 South Korea selected 31 follow-up projects and established an 'e-Government roadmap' to guide the nation and in its initiatives and coordinate the activities. The roadmap is aiming to complete the government-related projects by 2007 in a government-wide effort.

Increased private time

> **99% of phones sold in South Korea in 2005 were cameraphones**
> Source: MIC 2006

In many Asian countries it is still common to work to a six-day week. South Korea recently adopted the five-day working week, which then dramatically increased the available leisure time for its citizens. We should note that the working hours in the five work days typically run very long by European and American standards, and employees often still collect together for dinner together still discussing work related subjects. Nonetheless, that Saturdays and Sundays are now holidays for many, has greatly increased available spare time, and increased leisure activities.

One of the great beneficiaries has been the entertainment sector as the increase in spare time has in turn allowed a dramatic increase in leisure activities online, from gaming to music, TV etc. We discuss each of these in chapters of their own in this book.

B EDUCATED EMPLOYEES

As we started this chapter, a digital society has to start off by a digitally empowered population. This means the population needs to be well educated, and digitally connected. In fact all of the South Korean initiatives start off with excellence in its education. What is also surprising is how much individual South Korean citizens feel a need to improve themselves. Typical leisure time activities include learning better language skills such as English, though online training and on courses delivered via the cellphone etc. Similarly young Koreans learn computer programming skills, web design etc to give themselves a further advantage to their careers. The third most popular digital content on cellphones in South Korea, after music and gaming, is educational services.

Teachers

While it is important to educate the citizens and to provide IT technology to the schools, that alone is not enough. Also the teachers need to be trained. Here again, South Korea has taken a systematic approach, driven by government initiative, as the Korea Information Strategy Development Institute (KISDI) reported in its IT Industry Outlook Korea 2005 report:

> *By the end of 2000, all of the nation's 10,064 schools had finished LAN installation and Internet connections. Next steps will be to improve teacher's IT literacy, to develop new curriculum and teaching method using IT, and to produce new educational content using IT. Every teacher is required to participate in the duty training program on IT at least once every four years. The Information & Communication Officials Training Institute was established to provide IT education for government officials.*
> *IT Industry Outlook Korea 2005*, KISDI

After its citizens are well educated to understand the digital benefits, and after they have the connectivity needed, electronic information, available from anywhere, at anytime and from any device, has now become the main thrust of South Korean life. South Koreans have an active need to remain connected. So much so that many young adults now feel seriously agitated when they even think of not being connected or reachable.

Empowered employees

Being well educated upon graduation is of course not enough. Training the young citizens in schools and universities is relatively easy by managing the curriculum. A more challenging task is the existing adult population, which will also need to be retrained or further educated. The digital society is evolving so fast that continuous lifelong learning is required by members of society, to remain empowered and connected. The South Korean government is investing in an environment of citizen empowerment, that allows continuous e-learning, through infrastructure, courses, certifications, etc. South Korea's aim is to achieve the average of the OECD citizen digital know-how and the government is providing resources and incentives to achieve this ambitious goal. Part of the plan is to empower the local governments and giving them the access to national educational resources.

Labor

As digitalization started to have an effect on all sectors of society, the role of digital gatekeepers, the humans involved in developing and maintaining the system, became increasingly critical. The Sough Korean government actively promotes changes to enable this transition. For example, a change from the previous principle of lifetime employment within one employer and one skill, such as that in classic jobs of manufacturing, to the current career of multiple evolving skills and multiple employers, has increased flexibility but also the complexity of the labor market: This gives new opportunities for the ambitious.

For example various IT experts are now recognized as knowledge workers and as critical to the health of any business enterprise, and new cyber professions such as web masters are now fully respected. More exotic professions, such as Farmers of Digital Treasures inside multiplayer gaming environments, have also appeared. We discuss those in the gaming chapter. The growth in personal income, a more advanced level of education, awareness of the versatility of digital offerings have all added to an increased desire for quality in leisure and cultural life. These in turn introduce vast opportunities to offer digital services of entertainment, education, edu-tainment, etc.

C INNOVATIVE E-GOVERNMENT

A good indicator of how far South Korean society has advanced into the digital age is taxation, that mundane boring task. More than four out of five tax returns are processed online. The Home Tax Service system (HTS) was the starting point for using electronic tax filing services in 2002. Then in 2004 to further encourage participation HTS introduced a Tax Reimbursement System for Electronic Filing, which further contributed to enhancing the HTS system and take up in Korea.

Taxes online

As a result of the reimbursement system, the rate of electronic filing for VAT taxes currently amount to 73.6% as of January 2006, while that for corporate taxes and income taxes account for 96.9% and 81.2%, respectively. Perhaps even more amazing is that all the tax related activities with the government can be accomplished not only via the internet, but also all are available via the cellphone. Not limited to taxes, obviously, by 2004 the South Korean e-government was issuing 393 types of civil petition and processed 4,000 types of civil guides through the its digital e-government initiatives.

A central database today covers all residential registration information. Citizens are able to access the information in real time and check their data. The various other government databases are interconnected and a hierarchy of data integrity provides better accuracy than previous paper based bureaucratic means of maintaining citizen data. For example Korea's four major social insurance systems allow people to check their social insurance status online, to pay their social insurances, etc. If data changes in one, it is automatically updated on all four insurance systems such as a change in address, marital status, the birth of a new child, etc.

Procurement

Another area of government efficiency through e-government is logistics and procurement by the government. The Korean government processes the logistics figures of the public sectors on-line, which amount to 67 trillion won (approx $65 billion) worth of services per year. The transparency of this information helps in securing competitive bids for government contracts and adds to citizen participation and trust in its government.

The extent of fully digital processing of government matters is rapidly reaching very impressive rates. Back in 2003 the rate of electronic

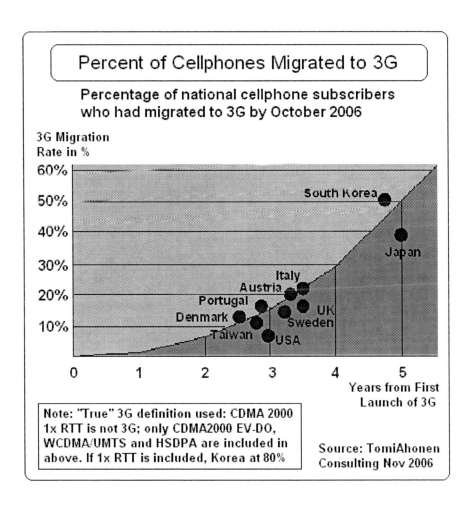

approval of proposals had reached 93.8%, and of all government documents, 88.9% were distributed electronically.

D LEGAL FRAMEWORK

The rapid pace of digitalization also requires changes in the laws. For example legally binding signatures to electronic documents were simply not covered in older contract law. The South Korean government also moved rapidly to adopt an advanced legal framework for its digital society. The Electronic Signature Law and the e-Commerce Law have been passed, clarifying the legal positions and removing ambiguities in dealing with the digital society.

Data protection

Another vital area is that of privacy in the digital age including data protection, which also is being constantly reviewed and made up-to-date. As a result, by July 2006, 35.7% of all South Korean banking transactions were already conducted online while more than half of South Korean citizens are active users of internet banking.

Pearl - No more regrettable calls to bosses when drunk. A new alcohol tester feature in a Samsung phone, as well as checking your ability to drive, allows users to block certain numbers - like your boss, your in-laws and ex-partners, etc. The alcohol tester disables calls to these numbers if the alcohol level is too high.

With the advent of user empowerment - user-created content - and the explicit support of the government, an amazing array of digital benefits have already been achieved. Ocean management systems have been deployed to monitor the various sea and sea life related data. National parks of Korea have various digital sensors to track matters from visitor numbers to forest fires. The museums, historical sites, cultural institutions etc are all adopting digital initiatives.

Coming back to the people

A full-scale knowledge-based society is not possible if only the government does it. The citizens must be activated and incentivized to participate in it. To fully join into it. The South Korean government has already provided information age training to 21 million people (43% of the total population) from the predictable such as civil servants, teachers, the disabled and elderly, to the surprising such as fishermen, farmers, housewives etc. Only by enabling the full society to gain from the digital benefits can the nation truly become a full Digital Korea.

Digital etiquette

New technical abilities may bring new opportunities and perhaps new freedoms, but with them come also the threats of chaos and confusion. Again South Korea takes a lead in guiding the development, understanding the role that government can play in ensuring that the transition does not cause enormous strains to society. For example, in February 2006, the government released a 256-page "*IT Ethics*" textbook for junior and high school students. Teachers are expected to spend 30 hours instructing from the textbook, whose chapters include "Healthy Mobile Phone Culture," and "Protecting Personal Privacy."

E E-HEALTH

The South Korean healthcare industry is naturally fully embracing e-government initiatives. Many of the typical initiatives known around the world are already commonplace, such as medical alerts, patient out care monitoring systems etc. Patient data, medication, medical history etc data is in digital form and available to treating doctors.

Diabetes phone

Some of the more advanced innovations are worth mentioning, as these capture some of the benefits from a comprehensive shift in society to broadband, high speed, wireless, digital community. In May 2004 for example, Healthpia a South Korean specialist medical technology company introduced the Diabetes Phone, which incorporated testing for blood sugar levels, and the over-the-air updates of the results for monitoring and alert services. The user would gain a history of test results, which could also be carried and displayed to any nurse or doctor involved in some other medical procedure or medication situation. The phone also includes fat analysis, alcohol levels, stress measurement etc.

Smart floors

In South Korean hospitals now pressure-sensitive floors are being introduced to help detect a fall of a patient and the "intelligent floor" to be able to alert and contact hospital staff to arrive to assist the fallen patient. These kinds of innovations are only the starting point. As doctors, nurses, pharmacists - and indeed the patients - are able to participate digitally in healthcare; innovations will be made continuously, including Healthcare Robots, which we mention in the Robotics chapter later in this book.

F LOCAL GOVERNMENT

Local government web sites are also very advanced. A good example is street maintenance. Let's say for example, that the street in front of you has repainted the white lines and accidentally made two of the parking places too small to fit a car. Today in South Korea the normal citizen could contact the local council digitally, search parking on the local government website, find the correct council department, then locate the hierarchy and even identify the actual government employee responsible for parking. The site would display a list of all of the responsibilities of that person, and most of all, the citizen could be able to contact that specific government person responsible for parking in that part of town, directly.

Collaboration is Key

The rapid pace of digitalization in South Korea has been achieved with an unprecedented level of collaboration in all aspects of society. Government has laid the foundation, supported the initiatives with legislation and government initiatives. When necessary, the government has also funded the early stages of the process. However, absolutely vital has been the close cooperation of South Korean industry and economy. The success of e-government can be contributed heavily to these forms of community service provider collaboration, which is a key ingredient in keeping Korea advanced in this field. This is similar to the mobile and internet consumer relationship with service providers maintaining the leading edge of next generation evolutions through digital community feedback and suggested improvements.

SMS in Politics

As digitalization spreads through all forms of government, it would be impossible to list all services, applications and innovations. One perhaps fascinating element was the SMS campaign by then candidate, now President Roh of South Korea in the elections of 2002. His campaign sent out 800,000 SMS text messages to spread his political message to the electorate. Its reach was considered the broadest political use of text messaging in electoral politics at the time. Since then SMS text messaging has become a staple in political activity in South Korea.

South Korea is also consistent in engaging and exporting its innovative ideas and best practices . The Ministry of Information and Communication itself has instrumented and been active on international

> **South Korean Youth SMS Usage**
>
> Survey of 1,100 South Koreans aged 14-19 on usage of SMS by the Korea Agency for Digital Opportunity and Promotion in 2005 found following results:
>
> | | |
> |---|---|
> | Send text messages in class | 40% |
> | Send text messages when bored | 40% |
> | Have received threatening text message | 20% |
> | Take cellphone to bathroom | 20% |
> | Have been bullied by text message | 10% |
>
> Source: Korea Agency for Digital Opportunity and Promotion November 2005

outward bound missions in UK, Germany, France, Portugal, Bulgaria, Chile, Azerbaijan, Mongolia, UAE, Tunisia, Algeria, El Salvador, Kazakhstan and India , and this is just a few examples within the first half of 2006 alone, many more hands on Ministerial events occur regularly.

G LAW ENFORCEMENT

South Koreans tend to be quite law-abiding and respectful of authority and their elders. Still, the advent of digital Korea has brought about both opportunities for improvements in law enforcement as well as new threats from digital criminals. New digital nuisances such as hacking into computers, malware and computer viruses, and the distribution of unsound information have also become a social issue. Furthermore, the protection of private information such as personal data and location data has also become important with the expansion of e-commerce.

> **20% of South Korean DMB cellphone owners watch digital TV in car**
> Source: Irdeto January 2007

A familiar theme for readers of our book, fighting cyber crime will also of course start already in school. One of the initiatives is the Cyber Crime Prevention and Correction Project in 2003 was a project to bring education about cyber crime to middle and high schools. For criminals actually convicted of cyber crimes, a cybercrime correction activity was set up with 22 probation offices in 2004, to attempt to correct disruptive cyber crime behavior.

Traffic Wardens snap Pictures

Traffic wardens in South Korea have been using digital cameras for years already to take pictures of illegally cars before they are towed away. The automated system recognizes the owner from the license plate on the car and the mobile phone number. The owner is sent an SMS text message with a link to the picture of the parking offense, together with the time stamp of when the offense happened, as well as the current location of the car and the fine to be paid. The digital proof reduces greatly the arguments about whether a fine should be paid or not.

Cyber security

As ever more of its society migrates to the digital space, the very digital integrity of South Korea becomes a top priority. Building the basis for a knowledge-based information society, The Korean Internet Security Center was established to ensure a reliable cyberspace and to effectively protect networks from cyber attacks. Advanced innovations are adopted and endorsed by the South Korean government such as electronic signature authentication systems that will allow users to use a single electronic signature in all digital activities. The digital integrity is also constantly upgraded such as through developing a next-generation standard encryption system.

The government constantly helps in the development of anti-spam technologies and launch campaigns that promote healthy moral standards, including the distribution of a spam mail prevention software.

Viruses and malware

The number of new viruses and the damage caused by them are increasing each year. While first only limited to the internet, now with digital convergence, virus threats appear also in cars, cellphones, digital TVs etc. The time it takes the hackers to penetrate the targeted system is getting shorter as hacking techniques continue to improve, and hacking itself is becoming ever more widespread. Often hacking tools and virus creation software are available on the net further complicating the task of keeping cyberspace secure. As such, the government has instituted countermeasures to viruses and malware by building a security system and establishing a cooperative system with the related law enforcement and other government agencies concerned in order to protect personal information and provide safe e-Government services.

Stolen phones

In South Korea there are no handset subsidies for cellphones, so there are no "one dollar phones". South Korean subscribers pay full street prices for their phones, and their replacement cycle is one of the fastest in the world. For the under 20 year olds, replacement cycles average under 11 months. Nevertheless, with the high cost of the phones and the clearly recognized value of phones, a stolen phone is usually reprogrammed and sold. While having a mobile phone stolen and others calls billed to a user's account may be costly, the potential for harm is much worse as both phones and thieves become ever more sophisticated now with downloads of videogames, MP3 songs, movies, etc. As the mobile wallet, credit card and banking access, as well as digital keys and passkeys to work become embedded onto the phone, the threats become even more pronounced and far-reaching.

Cyber terror

While individual attacks to persons, business enterprises and government departments by criminals are one part of cyber safety, a larger terrorist-initiated threat also emerges. As the society increasingly depends on secure digital interactions, centralized databases and vast commercial transactions, a terrorist attack on the digital foundations would place the most advanced digital societies like South Korea particularly at risk. Thus the government has initiated a Preparation of Countermeasures against Cyber Terror Background and Overview to ensure the stability of the administrative process and the protection of personal information also against a terrorist threat.

> **40% of South Koreans have created an avatar**
> Source: Neowiz 2006

A security initiative has been integrated into the e-Government Network, the dedicated information and communication network used by all government agencies since 1996. In 2002 the Integrated Security Control Center was built to share attack information among security personnel in real time and take collective measures for counteraction. The Center strengthened the monitoring of cyber terror against government agencies through 24 hour monitoring of hacking attacks against national information centers and monitoring the appearance and distribution of viruses as well as the effectiveness of countermeasures against them. Other similar government measures were also deployed.

Should an attack occur, the government expects to minimize recovery time from cyber attack through systematic review of the vulnerable points, risk analysis, asset and attack assessment, and security policy management. Furthermore, the risk management system will be overhauled to introduce reverse tracking of computer hacks and an automated patching system to vulnerable points to proactively respond to emergencies. A new threat response management system is being built based on the various cyber threats.

H WHERE DID THE MONEY COME FROM?

The digitalization of South Korea has been a huge investment. The work has been done rapidly and all of the country today benefits from the world's most advanced digital infrastructure. How was this achieved? Where did the money come from?

All part of the plan

Much of Korea's success as a ubiquitous information society can be attributed significantly to systematic investments and initiatives in information technologies and telecommunications. A good example is the money collected from mobile telecoms spectrum allocations. In some countries licenses were given out for almost free. In other countries the license fees were treated as a windfall benefit reducing the national deficit or simply added to the national budget as an income stream. In South Korea all revenues from telecoms spectrum licenses were strategically reinvested into the information technology and telecoms development. These considerable funds were directly channeled into the further investment often as the seed money for government initiatives from training to test centers to pilot projects to innovation awards etc. Thus a virtuous cycle was achieved.

However, all of the government funding of digital initiatives have been founded on a "launch initiative" principle. Only those concepts and ideas that would attract further private sector investment would be seeded with government money. This ensures that none of the government spending is wasted, but rather that they help jump-start areas of the industry. A good illustration of Korea's success over the last decade or so is the extent to which it has increased the level of spending on telecommunications as a percentage of GDP, where it has grown to be almost twice the global average by 2002.

The Future

For all of the impressive achievements in the past ten years, the next ten are promising to push South Korean digitalization even further ahead. There is a certain virtuous cycle that becomes a benefit to all and which in turn pulls all to participate. A good example is the U-City project led by KT Korea Telecom the fixed landline telecoms operator in South Korea.

The U-Cities (Ubiquitous Cities) concept is being built into five existing South Korean cities - Pusan, Songdo, Dongtan, Unjong and Hungdok - to turn them into futuristic places characterized by pervasive computing and ubiquitous communications. Many previously lifeless elements of the common city experience will be made fully digital with sensors and incorporated chips. Yun Hae-jong, the Vice President of KT in charge of the U-City project said: *"Revolutionizing traffic, healthcare, education and other lifestyles, the U-City project will create huge benefits for both people and related industries in Korea."*

> **Pearl - Village Information Center.** Rural digital development in South Korea has allowed village residents to share, collaborate and take pride in their region, as well as promote and sell local goods via e-commerce. Movies are now screened and world sporting events consumed together via the Village Information Center.

What extent of benefits will this bring is almost impossible to map. Some of the early visions include intelligent parking places to guide drivers to the nearest available parking slot; bridges to monitor themselves with sensors and report the need for maintenance; intelligent roads reporting on traffic loads; traffic lights re-reprogramming themselves when a crossing street has no waiting traffic, etc.

U School Example

The Daesin Elementary School in the southern Korea port city of Pusan is a good example of the digital future of elementary schools. As soon as the children arrive at a school, they put their student identification card on an electronic class board to register their attendance. This triggers an automatic text message to the parent, which says, "Your child has just arrived at school at the time of 8:35 a.m." The daily class schedules, homework notices, comments about school materials etc are all visible on an electronic board and accessible online and via cellphone.

With pervasive computing the classrooms make full use of computers in teaching, experiments, homework etc. Students can have their work immediately digitally checked. Mobile elements are included, such as the ability to virtually participate in a class trip taken by some other class or for example if the student is ill or unable to attend.

What about the Villagers?

While modern South Korea may be more known for a rapidly growing modern urban center in particular around Seoul, still 70% of the country is mountains and dense forests in a country the size of Portugal or the US State of Indiana. Bringing digital benefits beyond the cities is another vital interest for the South Korean government.

For the rural areas, with small villages, South Korea introduced a digital initiative called Information Network Village Project (INVIL).Through INVIL the government hoped they could bridge the digital divide for rural areas. The project aimed to improve the penetration of computers, usage of the internet, and delivery of the benefits of the digital world to the rural areas, such as access to culture, entertainment etc. There are obvious rural area benefits such as information about health uses, prices of agricultural items etc.

As the infrastructure was put into place by INVIL, numerous local digital initiatives appeared spontaneously. By 2006 a total of 523 club activities, 1,421 personal web sites and 1,694 blogs have been launched in rural areas covered by INVIL. 62.4% of participants said they had been able to strengthen their bonds with family and neighbors. The program resulted in strong increase in internet use, where 52.4% of the INVIL participants achieved above average Korean levels of internet use.

Conclusion

Taking a comprehensive view to digitalization and studying the Korean model can give insights to other countries and governments around the world, on to how achieve successful mobile information services. The 2004 survey of municipal digital governance of cities by Rutgers University found unsurprisingly that Seoul was the worlds' top-performing city for digital governance. In various surveys around the world South Korean leadership in e-government initiatives is being recognized. In that way Digital Korea can act both as an example and as a case study. Certainly most of the initiatives from the Korean experience can be replicated in other countries.

Case Study 6
Ohmy News

Ohmy News is one of the inventors and innovators of the concept of citizen journalism. Founded by Oh Yeon-ho who still acts as Editor in Chief, Ohmy News has rapidly grown and today features over 50,000 citizen contributors to the daily online newspaper with about 1.2 million daily readers in South Korea, an English Language foreign edition, also written by citizen journalists, and the first expansion overseas with the launch of Ohmy News Japan in 2006.

The Ohmy News model includes full crediting of all citizen journalist contributions whether contributing to an article or supplying a picture etc. The stories are also rated by readers and best contributors can be virtually "tipped" by readers, which in turn is converted into money rewards to the writers. What seems to be a common theme, most contributors to Ohmy News are more motivated by seeing their name in print, than the monetary rewards of their contributions.

Ohmy News is not a purely amateur organization. To assure high quality content for the daily online newspaper, Ohmy News employs a large staff of professional editors, who are particularly competent at supporting the citizen journalists in delivering high quality editorial material to the paper. And Mr Oh described the difficulty of attracting competent citizen journalists when he spoke to 2nd International Citizen Reporters Forum in 2006:

> Not everyone can write a news story. Only those citizen reporters who are passionately committed to social change and reporting make our project possible. The main reason that citizen journalism has not grown and spread more rapidly is the difficult task of finding and organizing these passionate citizen reporters in waiting.
> Oh Yeon-ho, Founder and Editor of Ohmy News

While many in the industrialized world might initially doubt the quality of a citizen-generated newspaper "but they are not professional journalists, how can they get the story right", the actual findings from the performance of Ohmy News are quite different. As they have literally an army of reporters, Ohmy News has a much larger staff to collect the elements to a story. That in turn produces more initial material - and indeed more editing work - but helps compensate for any lack in traditional skills of investigative journalism. The reader surveys and opinion polls in South Korea have seen a steady rise in the reputation of Ohmy News, with the latest findings having Ohmy News rated fourth most trusted news source in South Korea, ahead of several TV stations, most other newspapers and all radio and other web news.

As the popularity of Ohmy News has also grown, it is also becoming the preferred media for those advocating the new. South Korea's new President, Roh Moo-hyun granted his first press interview to Ohmy News in February 2004. And as this journalistic innovation gains credibility in South Korea, word spreads around the world. Similar citizen journalist online papers have been set up in countries from Japan to Canada. British media and advertising guru, Alan Moore the CEO of SMLXL Small Medium Large Xtralarge, explained the relevance of Ohmy News in interviews for this book, and we want to end this case study with his thought:

> OhMyNews is a key contribution to the development of news journalism, globally. In a world where trust is in short supply, news media companies are realising that people can and do want to make a meaningful contribution to our societies. That they have become distrusting to incumbent news organisations. Oh My News is a beacon of what the future could hold for us.
>
> Alan Moore, CEO SMLXL

Chapter VIII
Machine Telematics

The Intelligent Automobile

Image courtesy *IT Korea Journal*

> *"Automotive telematics is moving from something that's directed inward, toward monitoring the conditions of the car, toward a system that's also monitoring the environment around the car."*
> **Mike Liebhold, Institute for the Future**

VIII
Machine Telematics
The Intelligent Automobile

The intelligent car, the connected car. The world has about 850 million automobiles in use and the Trillion dollar automobile industry has been looking at the computing and connectivity dimensions to build value to their cars. The car industry talks about telematics. Telematics is actually a much larger field than just connecting cars, and includes portable GPS devices and short-range RFID devices etc. In this chapter we explore those matters, but we start with how far the converging automobile, computer and communication industries of South Korea have come in deploying car telematics solutions. Fasten your seatbelts...

A THE NEXT INTERNET IS THE CAR?

The first internet was on the personal computers. The second internet is now moving to cellphones, as for example the majority of internet access in Japan, China and South Korea already happens from cellphones, no longer from personal computers. Google has said that the future of the internet is on cellphones. However, if you already live in that future where the internet has migrated from the PC to phone, is it perhaps moving further, *beyond* cellphones?

Car and web

The South Korean Government has argued that the third internet arena will be the automobile. While perhaps not as big in terms of total users as the cellphone based mobile internet, or the location-fixed computer based broadband internet, looking at South Korea today, a new internet consumption environment seems to be forming inside and around the automobile. Not one so much working on surfing and e-mail as the classic PC based internet; nor one working on alerts and SMS as the mobile, i.e. cellphone based portable internet; the car based internet would feature location-based real time maps and navigation guidance as its early applications of most relevance.

South Korea already features numerous cars with advanced telematics systems as factory built solutions such as the luxury end of Hyundai Kia such as the Grandeur XG, the EF Sonata and the Regal. Also for example Renault has the Samsung telematics systems in its vehicles sold in South Korea.

Each of the three mobile operators, SK Telecom, KTF and LG Telecom offer telematics solutions with the fixed wireline carrier KT - Korea Telecom also providing telematics solutions. Numerous third party solution providers are building business out of the rapidly growing telematics area for automobiles. Let us start with examining the technology to pinpoint where we are, GPS.

Where am I?

The satellite based GPS (Global Positioning System) has been available for commercial use for many years already. Early commercial applications for this technology initially developed for the military, have been deployed for tracking fleets of cars and providing emergency assistance.

The South Korean pioneer in GPS solutions, Axess Telecom was recently renamed Ubistar after acquiring telecoms vendor Intelinx. Ubistar offers GPS modules to Samsung, Hyundai and Kia, and also offers its own services under the Roadmate and Roadfree brands. The service offered by Ubistar goes beyond the location identification of the vehicle as Ubistar provides a traffic database, which is linked to location that is updated weekly. Ubistar provides over half of all GPS modules sold in South Korea today and has set as its target to attract 500,000 car owners in South Korea as its customers. Ubistar is now developing GPS modules to cellphones.

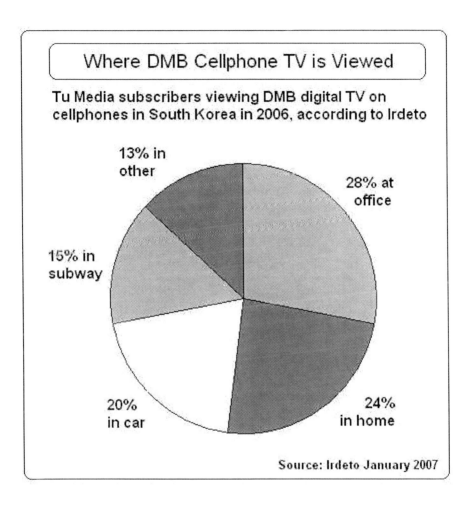

LG and GM

LG Electronics was designated as a telematics device provider by GM (General Motors) in America in November 2003. LG Electronics is developing new products that will meet needs beyond South Korea aiming for the Japanese and North American car markets as well. In September 2004, LG Electronics and Hyundai Motors set out to build a research center for technologies in vehicle multimedia services and solutions. LG Electronics looks beyond simply the location of the car and services around location. LG

Electronics is targeting car infotainment as the direction of its telematics business and hopes to deliver experiences to the car similar to what we have in home entertainment. Of course the service also includes GPS and services based on location and geography.

Mozen

At the top end of the South Korean telematics offerings today is Mozen. Mozen is a complete one-stop-shop solution for all connectivity, location, information and entertainment services one might want in a car. Mozen includes real time traffic data, route guidance, remote diagnostics of the vehicle, TV content and internet access. It includes customized consierge services, which can book restaurants etc on behalf of the driver. It is quite an expensive solution with the Mozen terminal costing 6 million won (6,400 US dollars) and the subscription price of 20,000 won per month (21 US dollars)

11% of Americans use eMail on cellphone
Source: m:Metrics March 2006

The high purchase cost of the Mozen terminal has led to a less costly device for navigation only with a price at about 1 million won (1,070 US dollars). It is not a bare-bones device, as even that includes an MP3 player, VCD videodisk player and wireless internet access. The device is sold through Hyundai as the HNS-5000 and also by third party providers in South Korea such as Canas, Audiosound and Thinkware Autonet.

TV in the car?

It might seem strange to Western readers to consider a TV receiver or video player in a car. Surely it must be disabled for the driver? But in Seoul's notorious traffic gridlocks, it is not uncommon to see car drivers with a portable TV set on the dash board to watch TV while stuck in the traffic jam. The early user research of where DMB was viewed suggested that the automobile was the fourth most common place where this new digital TV was being consumed. Most of that was by the drivers, not the passengers in the car. As the entertainment services evolve and become more user-friendly, they also start to adopt electronic program guides and voice based controls which then further enhance the entertainment opportunities inside the car.

These are not the only car telematics services and products in South Korea. The launch of DMB digital TV services for portable devices, as we discussed in the TV chapter earlier in this book, has brought a fresh round of competition in the car telematics space who are now also integrating touch-screen technologies and TFT screens. Several other providers including Fine Digital and Jaty Electronics also provide technologies such as FineDrive (Fine Digital) and Gilbeot (Jaty) which have prices in the sub 500,000 won (500 US dollar) market segment.

B PORTABLE DEVICES

Aiming for services and utility beyond simply the car, Carpoint has worked

> **21% of South Koreans use eMail on cellphone**
> Source: NIDA September 2005

with South Korean mobile operator KTF to develop portable car telematics devices and services. Its Xroad service has a portable telematics terminal, K-Ways Wide. Carpoint offers tracking services under its NBox solution. The services include basic location and guidance such as maps but aiming for this segment, they also offer alerts, a messaging function and alert, internet access, as well as entertainment functions such as gaming and music and are easy to use with touch screen technology. Carpoint is also deploying voice activated calling, DMB TV reception and beacon based traffic information. It is also developing solutions to work on cellphones.

Saving lives

The obvious application of telematics, real time traffic information and optimized navigation is of course for the emergency services in saving lives. South Korean fire fighting and rescue services were among the first to deploy telematics solutions to get faster to the site of the emergency.

Telematics Services and phones

Location based services have been launched in most countries during this decade, mostly to modest success at best. The services have suffered from a lack of accuracy by cellular networks and a lack of compelling services. Still many telecoms analysts seem optimistic on "LBS" or Location-Based Services and hope that the introduction of GPS technology will finally crack this nut.

In South Korea Samsung launched the SPHS 1100 as its first Telematics Phone with inbuilt GPS, which was also the first telematics phone, released in South Korea. It ships with preinstalled national maps of South Korea, which makes the map utility greater than previous mapping and guiding services on cellphones, which mostly rely on guidance maps that are downloaded off the air, to delay and cost for the end-user. The phone offers spoken guidance so it can be used in the car as the navigator. The rest of the phone is typical to the high requirements of the South Korean market with MP3 player, camera, high-resolution screen, internet access etc. For drivers the services includes an emergency button to get direct contact with towing and repair services. On the KTF network the Samsung SPHS 1100 is integrated with the KWays telematics service of KTF as well as its K-Bank mobile commerce solution.

Mobile operators also in game

Naturally the innovative South Korean mobile operators are all into the telematics opportunity; after all the original name for the cellphone/mobile phone was the carphone. The three South Korean carriers (mobile operators) are engaged in a heavy competition for lead in not only local innovation but also leadership in the industry on the world scale. Each of the operators offers "LBS" Location Based Services and telematics solutions usually delivered via the handset. A survey by NIDA (National Internet Development Agency) of South Korea in September 2005 revealed that location-based and telematics services were already the ninth most popular category of mobile internet content in South Korea, used by 8% of the subscribers.

Competition benefits

A critical component of building the digital ecosystem for telematics is rich competition in the various technology platforms. Often overlooked, this has been a key element in how South Korea has achieved its lead in all digital

Chapter 8 - Machine Telematics

areas and it helps explain why so many different players are active for example in the telematics space. One of Asia's leading IT/telecoms analysts, Walter Adamson, the founder of Digital Investor in Australia, was perhaps the first to point this out in his insightful White Paper on South Korea, in 2003:

> *One of the most far-sighted decisions was the establishment of a framework for facilities-based competition. The Government saw poor progress on the part of the incumbents sharing their networks, as 'service-based competition', and so eliminated entry barriers for new network operators. They set up 'facilities-based competition' through new network operators, as opposed to 'service-based competition' through sharing the same network. This single policy set the scene for an enormous build-out of capacity, and very active innovation in technology, equipment and service. In contrast, the network-sharing model, proclaimed by many other countries, provides the incumbents with no incentive for technology and equipment innovations.*
>
> Walter Adamson *"How Korea Built the Broadband that America Wanted"* White Paper, 2003

SK Telecom, the largest of the South Korean wireless carriers (mobile operators) is a good example of this. SK Telecom offers a vast range of wireless services on its Nate service, South Korea has become the second country in the world after Japan where more of the total wireless data revenues come now from non-messaging services, and SK Telecom's Nate is often compared to NTT DoCoMo's i-Mode as the innovation leader in this space worldwide.

For many years already Nate has included a growing range of telematics services. These mostly do not require GPS positioning, but rather work on a triangulation technology on the CDMA network to give reasonably accurate positioning for most typical location and telematics applications, such as showing a map of where the subscriber is currently. The service is now starting to integrate GPS data as GPS enabled handsets are appearing onto the market.

Some of the services that have been launched in South Korea that combine wireless services of the cellular network and are used on cellphones and PDAs include Inavi by Thinkware; SpeedNavi by Mando Map and Soft; Enjoymap Moti by Navtech and PocketNavi by Citus. The early adoption of navigation, location and telematics services on cellphones had also suffered from the relatively small screens of early smartphones, but with current phones featuring ever-larger screens, these are also more suitable for use in displaying maps. Industry experts hope that this will enhance the appeal of these services.

How about pagers

Not to be left out of the telematics and navigation opportunity, the South Korean pager (i.e. "beeper") service provider RealTelecom has introduced its Real Traffic service based on wireless data but not including the voice services that the cellular providers offer. Real Traffic powers solutions such as Hyundai's Autonet and telematics data services by the broadcaster KBS and the cellular network operator KTF. The Real Traffic service provides news, weather as well as traffic news in basic text format. Obviously the service does not have GPS nor maps, but provides valuable traffic information at a small fraction of the cost.

C TRAFFIC AND PARKING

Part of the car-related telematics industry is real-time traffic data in digital form. Sensors and systems to collect real time data on traffic have been deployed in various South Korean cities and major routes and real-time traffic information has been available for cellphones since 2002 in South Korea.

> **Pearl - Best price for petrol by cellphone.** If you are running low on your fuel on your car, you can dial up a service on your cellphone and find out which of the nearby petrol stations offers the lowest price today. The service covers a radius of about a mile around you.

In many countries it is already common to pay for parking by cellphone. Certainly it is a major improvement to coin-operated parking meters when you do not know exactly how long your meeting might last, and it may be raining and you might not have the right change in your pocket. The cellphone based mobile parking payments solutions solve all these problems and were first introduced in Norway in 2000 and already today for example in Croatia over half of all parking fees are paid by cellphone.

Beyond paying for parking

In South Korea too it is possible to pay for parking by cellphone. Of course. But now the holy grail is to evolve the mobile parking solution to be more useful. The perfect parking solution would find the nearest available parking slot. That is the real killer application for drivers and car telematics. Several pilot projects are ongoing in South Korea to embed parking slots with sensors, for example a simple light sensor, to detect if the slot has a car above it or not. Then this data is collected into a central parking database managing all legal parking slots of a given township for example. This data would then be available for the driver, helping him to locate the nearest empty parking place. Some smaller systems like that have been introduced in major parking buildings for example, but nothing like it on a township wide scale. Not yet. Expect this to be launched soon in South Korea. The technology to deploy this system exists already.

D RFID SYSTEMS

A big growth opportunity in the digital space in South Korea are technologies utilizing RFID (Radio Frequency IDentification) the very low cost and short range wireless data system. Typically RFID has a range of one meter (three feet). There are various commercial applications in home appliances, retail, food, clothing etc, but now telematics companies are exploring opportunities in RFID. Moreover, as is so typical for South Korean industry, the government has taken a pro-active role in promoting RFID and supporting the fledgling industry. The MIC (Ministry of Information and Communication) has worked with the Korea RFID/USN Association and RFID has been brought to the attention of ministries from commerce and agriculture to aviation and defense. RFID is seen as a critical technology for next generation solutions in retail, with Shinsegae department stores and E-Mart convenience stores in South Korea developing solutions.

Two of the early companies developing RFID for car telematics are Hyundai Motors and Ssangyong Motors who have introduced RFID technologies into the manufacturing process. Car manufacturing plant experts suggest RFID will become a significant part of the painting management on the assembly line as well as documenting greater details of production history for management accounting.

RFID Applications

Numerous RFID services exist in the world. In South Korea they are rapidly being deployed partly due to the low cost of the technology and its great added utility allowing digital handling of non-digital products and services, which occupy the "bricks and mortar" real space of the world. The first services deployed in South Korea were in the parcel delivery and postal services where an RFID tag is attached to the parcels being delivered. These will then emit a unique digital signature allowing them to be sorted automatically. The accuracy is far greater than reading addresses optically, even when most parcels have addresses written by machines (computer printed). RFID can even bring greater efficiencies than using barcodes as the ID can be read with much greater speed on moving objects.

The next application came from the warehousing and retail storage management in logistics and distribution. RFID allows immediate digital reading of a stock of a given product category, allowing accurate rotation of stock that is going to be stale or obsolete, critical in perishable goods management for example, such as fresh foods, flowers etc. Specialized stock-management RFID reader industrial handsets exist which integrate with the warehouse logistics IT systems.

How is your casino doing?

Embedded RFID inside casino chips. A fascinating specialist application for RFID is casino chips. All casinos use custom money that is only used inside

---

Location-Based Services that identify the owner's position are used by 8% of cellphone owners in South Korea
Source: NIDA September 2005

---

the casino, and needs to be exchanged into the regular currency of the country. Casino chips therefore are custom built and tend to be optimized for casino use - being larger than coins, but usually of uniform diameter, often with ridges to allow easy stacking. Casinos are built on managing the money flow, and the more accuracy they can achieve in understanding which roulette wheel or poker table or slot machine is losing money, the better the casino can optimize its business.

Enter the RFID embedded casino chip. The latest innovation in casino money management is embedding casino chips with RFID. Now every casino chip can be individually tracked and monitored. With clever casino management software, it becomes possible to have real-time information on how every croupier is doing and how every gambler is doing. When a given visitor is on a lucky streak, the casino will know immediately, and may want to give complimentary drinks etc (perhaps also to slow the winning streak). The same is true of criminal activity. Numerous swindlers attack casinos regularly with the latest schemes to try to beat the casinos at their various games of chance. If the casino knows in real time how each gambler is doing based on his chip supply, the casino can easily spot the exceptions to the rule, which customers are clearly behaving against the statistical norms and therefore warrant close inspection: these might be cheating.

Mobile RFID

An area that shows great promise for new applications and services is the merger of cellular technology i.e. the cellphone and RFID. What is called Mobile RFID, is integrating the RFID reader into the cellphone and allowing integrated services with the cellphone the device that enables these. Some of the opportunities are using the cellphone as the RFID terminal, for anything from MP3 playback, gaming, video content, picture transfer and printing and of course, telematics. The particular strength of integrating RFID into the handset is that all economically viable people on the planet already carry a cellphone, and differing from technologies such as 3G and WiFi, adding RFID is a low cost component of modest drain on cellphone performance, size, weight and battery. The Chairman of the Mobile RFID Forum, Hyeok-Jae Lee has said:

> "*RFID will provide a solution for the worldwide IT industry, which has entered a cycle of stagnation. Korea's mobile RFID will lead the world and become the core of the next-generation IT industry.*"
> Hyeok-Jae Lee, Chairman Mobile RFID Forum

Handsets with RFID have not yet been released commercially, but considering how rapidly recent technical innovations such as DMB TV and

2D barcodes have spread in South Korea; odds are that RFID will become a significant technology in the near future.

Future applications of mobile RFID

It is nearly impossible to guess what will be the major areas where mobile RFID can succeed. It is too early for this type of technology. The South Korean industry has suggested several application areas, which show promise.

Mobile advertisements via mobile RFID. An obvious early application for the technology is embedding the RFID tag to various posters, leaflets, brochures and other advertising. The RFID will include a link to a messaging phone number. When the user with the handset scans the RFID, the phone will retrieve the advertisement, which can be an SMS or MMS message or perhaps a videoclip of an advertisement of the product. The RFID can also offer the web address for the product with more information. Naturally, discount coupons etc can be distributed in this manner. The contact to the RFID would be free, as very short-range radio communication. Any resulting traffic on the cellular network (such as the phone calling up a message that is sent to the phone) would need to be charged. These should of course be free to the end-user, with the advertiser covering the cost of the message (advertisement) delivery to the handset.

Other applications include opinion polls and surveys. An RFID tag can be integrated into a poster about the survey. Any passers-by could point their phone at the poster, which would launch the survey and collect the data via RFID and the user's cellphone. These kinds of opinion surveys could also include rewards by the polling organization that can be provided directly to the cellphone account of the user taking the survey.

Mobile RFID can be used in the information services around public transportation, such as being at bus stops and on subway maps etc. The service could guide the user into taking the optimal route, or advise for example on where wheelchair access is possible etc. The technology is also suited for various community bulletin boards and providing tourist information. The beauty in services around mobile RFID is that almost unlimited further information can be provided through the cellular network, without users typing in cumbersome website addresses etc, when the links are provided through RFID to the cellphone.

On business applications RFIDs are also expected soon to be embedded into business cards, as 2D barcodes are now becoming the standard in South Korea and Japan. Applications have also been suggested in areas of pets, such as embedding the pet owner information to the collars of the dogs and cats - or even surgically embedding the RFID chip into the pet, to recover lost pets and to track animals etc. Similar solutions are being deployed for cattle management for example.

Even shorter-range radio technology is being introduced, called NFC (Near Field Communications) where the total receiving range is measured in centimeters (inches). These again have a less sensitive, smaller and less costly radio unit, and the future almost definitely includes solutions built on NFC.

Case Study 7
3G Traffic Cam

Merging the 3G cellphone and live traffic cameras has the appeal of a science fiction style futuristic application. As 3G networks went online around the world, numerous live cam applications have been launched, from traffic cams to surf cams. In South Korea as the usage was large enough, the service also soon evolved past simply the clever.

While it is easy to understand the premise of a traffic cam "wouldn't it be nice to have your 3G cellphone access all the traffic cams on our route and see where the bottlenecks are." Yes, the concept is easy. But the actual service in use, is not that convenient. Viewing every traffic camera scene, for 10 to 15 seconds at a time, to determine if that truck is moving at about the same speed today at this camera, as the cars moved yesterday, is actually very cumbersome. It works, but it is not fast and efficient.

The South Korean innovation to the 3G cellphone views to the traffic cam is a speed statistic service. The driver sets up his preferred route(s) and the network will give average speeds on every legs of the route (by each camera location), in real time.

My Route	
Intersection 4	24.3 mph
Intersection 7	22.7 mph
Pine Street	11.4 mph
Main Street	22.6 mph
Park Avenue	21.8 mph

With this kind of summary statistics, it is very clear to see on the screen that there is a problem on Pine Street and that camera is worth looking at.

> Now rather than scrolling across a dozen camera views, on simply one screen with the average speeds listed, the driver can spot the traffic bottlenecks immediately, on one glance. The cool idea has evolved and matured, and many drivers never leave home without looking at this statistic. A picture may be worth a thousand words, but a statistic is *faster* than a moving picture. And 3G is all about speed.

Chapter IX
Mobile Music

What After the iPod?

Image courtesy *IT Korea Journal*

"MP3 playing phones will kill iPods."
Bill Gates

IX
Mobile Music
What after the iPod?

A lot of the IT industry and global business press have been excited about the success of the Apple iPod and its iTunes music service. iPod was launched in the last quarter of 2001. What is much less well known is that two years later, in the summer of 2003, full-track music downloads were introduced for cellphones. Not as ringing tones, but full songs, in MP3 format. This innovation happened in South Korea. In only two years by 2005, music sold to cellphones worldwide exceeded the dollar value of music sold to iTunes. By the summer of 2006 musicphones were outselling iPods at a ratio of 6 to 1 and even Apple saw the writing on the wall by announcing the iPhone in January 2007. While it was a huge innovation for the music industry, the sale of MP3 music to cellphones is only the tip of the iceberg for the digital music experience in South Korea.

A RINGING TONES

Downloadable ringing tones, the most simplistic form of music for cellphones - and also the launch of value-add services for mobile, hence the dawn of the mobile internet, was invented in Finland in 1998 by Saunalahti (renamed Jippii Group and now part of the Elisa Group). Ringing tones were regularly pooh-poohed by the serious telecoms experts who at the time were planning the more advanced 3G future for cellphones. Meanwhile against the conventional wisdom held by almost all analysts, ringing tones went from strength to strength, each year outselling the previous sales records. In 2005 when iTunes worldwide generated 440 million dollars of music sales, one ringing tone alone, Crazy Frog, earned 500 million dollars. In 2006 the value of all basic ringing tones sold worldwide was over 6.5 billion dollars, not even counting the more advanced ringing tones such as Truetones and many other variants of the concept such as "ringback" (i.e. waiting) tones.

Not just the young

Many continue to think ringing tones are a silly concept only appealing to kids. But music is a universal passion. All of us have some favorite music. While it may not be the latest rap tune, middle-aged people may reminisce fondly about music by Blondie or David Bowie or Queen or Abba. People at retiring age may prefer Elvis, the Beatles, early Rolling Stones and the Beach Boys. Others will find their music tastes change, from pop music to jazz, classical music or show tunes for example. However, all of us have favorite music. In addition, we can see it in South Korea, even with ringing tones.

Rather than resist the customer preferences and demand in fostering innovation, the South Korean IT and telecoms industry is eager to supply all the services their customers want. Thus all carriers were very happy to provide a full range of ringing tones, and the multitude of variants and developments to them. Today, the NIDA (National Internet Development Agency) of South Korea reports that ringing tones are used by 97% of South Korean cellphone owners. Obviously the whole population can enjoy ringing tones. Yes, young people attracted to pop songs that are on the charts, will replace their ringing tones much more frequently than the older parts of the population, but there is no denying the appeal of music on the phone. Moreover, it starts with the humble ringing tone.

Waiting tones (Ringback tones)

An interesting twist on the ringing tone is the waiting tone. This extremely simple service concept was developed by WiderThan in South Korea, where the first commercial service was called Color Ring, and the worldwide telecoms technical term - and unfortunately often this confusing term is the commercial name for the service - called ringback tone. We prefer to call it the Waiting Tone, which makes more sense to the cellphone owner. The service does not ring on the "back" of the phone, nor does the service "call back" or ring back the missed calls, nor is it playing the song backwards, etc.

Mobile Internet Revenues Globally 2006

Cellphone downloaded content revenues worldwide in 2006 by categories in millions of dollars according to Informa (excludes SMS text messaging revenues)

Music (including ringtones)	$6,819 M
Games	$4,018 M
Video	$2,517 M
Gambling	$2,135 M
Adult	$1,255 M
Other	$4,647 M
TOTAL content revenues	**$21,392 M**

Source: Informa Mobile Industry Outlook 2006 Report

Ringback tone is an unfortunate name for the service, where Waiting Tone makes easily a lot more sense.

Waiting Tones (Ringback Tones) are music that replaces the "brrr-brrr" network buzzing noises you hear when you wait for someone to answer their phone. Think of it as music-on-hold, *before* you are put on hold. Therefore, when you call me, I can use my ringing tone to play music to myself and people near me. But when you call me, I can use the Waiting Tone to play music for you, before I answer the phone (or my voicemail service kicks in).

Soon bigger than ringing tones

Exactly like with ringing tones, the subscriber purchases the song he wants, and also like ringing tones, the subscriber can change the songs as the music preferences change. However, separately from ringing tones the subscriber also pays a monthly fee, and very significantly as this is a core network signaling solution, there is no competition for service providers. The service is not dependent on the types or abilities of the handsets and much higher quality music can be delivered as waiting tones. All the money is divided between the telecoms operator and the recording industry.

In just 8 months from launch, SK Telecom was earning more from waiting tones than from ringing tones. In a year over a third of its subscriber base was using the service. Both of its domestic rivals, LG Telecom and KTF have launched waiting tones. In South Korea alone the service is delivered over 150 M dollars in its first full year. Then as the secret was out, Waiting Tones were rapidly launched in Taiwan, China, Israel, Singapore and Japan

---

7% of Germans buy ringing tones
Source: m:Metrics March 2006

---

with Europe and America following later. Most of the early countries report solid Waiting Tones sales and soon revenues from Waiting Tones exceed that of ringing tones. Worldwide by 2006 the sales of Waiting Tones generated over a billion dollars of revenues. Meanwhile WiderThan was bought up by Real Networks to boost their knowhow in the mobile and converged music space.

B MP3 FILES

In June of 2003, as a trial, Sony Music recording artist Ricky Martin pre-released several tracks from his newest album as MP3 files sold directly to cellphone owners six days before the album was released. As only South Korea had the penetration of suitable cellphones at the time, this trial was limited to South Korea. In six days Ricky Martin sold over 100,000 songs. Since then music consumption has grown explosively. Two years later, by

September 2005, 45% of all South Korean cellphone owners download full track MP3 songs to their cellphones, as reported by NIDA. For contrast, five years after launch in America, the total Apple iPod penetration was about 15% of the population, and users of iTunes downloads obviously less than the total installed base of iPod users, as the majority of iPod music was transferred from music CD collections rather than downloaded from iTunes music store (Apple does not provide the exact breakdown).

By the summer of 2005 even on the smallest of the three mobile operators in South Korea, LG Telecom, 5.5 million songs were downloaded every month and generate 10 million dollars of revenues just on that service every month (annually that works out to 66 million songs and 120 million dollars of revenues). Again for contrast, Apple, with its iTunes service sold 31 million tracks worldwide in its first 24 months. Songs as MP3 files are reasonably priced in South Korea, nothing like the 2.50 US dollar prices on some networks. A single MP3 full-track download of a South Korean artist to a cellphone typically costs about 45 cents per song, and MP3 full-track

---

97% of South Koreans buy ringing tones
Source NIDA September 2005

---

downloads by Western artists cost about 80 cents to download directly to the phone.

Musicphones

South Korean phone manufacturers are among the leaders in developing the most advanced musicphones. Samsung was first to release a cellphone with more capacity than the iPod Nano, at 5GB, and has upgraded rapidly the top end of its musicphone line to 6GB, then 8GB and today already 10GB of music storage. Not to be outdone, LG released its Chocolate music phone, which in Europe has become the best-selling cellphone model of all time, according to Europe's largest phone retailer chain, Carphone Warehouse. In our interviews for this book, Woo-Jae Lee, the General Manager of Infraware of South Korea explained:

"There is a compelling proposition for music on phones when compared with online services like iTunes/iPod. With integrated music players on phones and the 3G high-speed wireless access, users do not need to separately download files to a PC and then transfer songs from PC to music player. The solution is easier and faster on mobile."

Woo-Jae Lee, GM of Infraware

The worldwide music industry knows this as well, and by early 2006 senior executives from each of the four giant music labels - Warner Music, EMI, Sony BMG and Universal Music had come out during 2005-2006 all echoing the same theme: the natural future home of music will be on cellphones with direct downloads of full track songs. Not stand-alone MP3 players like iPods nor online music stores like iTunes nor separate music CD sales whether in stores or online:

Warner Music's Chairman and CEO, Edgar Bronfman, said, *"Wireless will become the most formidable music platform on the planet."*

EMI Vice President of Digital Development Ted Cohen admitted that the cellphone will win out over stand-alone music players as he put it *"The cellphone will become the digital music player of choice".*

The General Manager of Universal Music, Rio Caraeff says it like this, *"Music is inherently mobile and something you enjoy on the go."*

While Sony BMG Senior Vice President of Digital Business, JJ Rosen says it like this, *"Everyone likes music, and everyone has a cellphone."*

Then finally even Apple's Steve Jobs joined in at the announcement of the iPhone in January at Macworld, Steve Jobs said that the future of MP3 players belonged to musicphones.

Apple's iPhone, not revolutionary in Korea

While the world was very amazed at Apple's unveiling of the iPhone in early 2007, a very similar, large screen, touch-screen phone of almost identical form factor, had *already been revealed* by South Korean manufacturer, LG, together with fashion brand, Prada. The LG "Prada" phone had in fact won an industrial design award before the iPhone was even announced.

Where the West Coast of America is waking up to fashionable cellphones and with Apple's considerable marketing presence and leadership, in South Korea this type of innovation is commonplace. The iPhone, at least by its superficial appearance, was not seen as revolutionary at all. A clever tweak of the concept, perhaps, and time will tell what the user interface and "multi-touch" screen and various sensors will do, but where the Western press was overwhelmed by the iPhone launch, the Asian reviews were quite underwhelmed. Perhaps that is why the iPhone will launch in the USA first, and in Europe before Asia. It is an advanced smartphone for a laggard market.

Melon Music

We discuss Melon in the case study at the back of this chapter, so we will introduce Melon just briefly here. Melon is perhaps the most advanced music service in the world, being the first fully convergent music portal. It features over-the-air music downloads, real-time music listening, and the whole service is available on cellphones, personal computers and stand-alone MP3 players. Developed by WiderThan and being part of the biggest carrier, SK Telecom, Melon also reaches the most of the South Korean users.

Will not stop at sound of music

In addition, while many will be bewildered by the rapid pace of cannibalization of the iPod market by music-playing cellphones, a much bigger revolution will be happening simultaneously around music. Just like twenty years ago when MTV revolutionized the way music was consumed, and music artists and fans discovered the music video, this will very soon become the preferred means of consuming music on the go. Already today in South Korea and several other advanced mobile markets like Japan and the UK, music videos form a major part of the total music sales revenues on cellphones.

C MAKE YOUR OWN MUSIC

But wait. The phone is not only a consumption device like the iPod. The phone is far superior. It is more like a pocket Mac. A pocket computer, fully capable in all kinds of creative applications. Want to compose a ringing tone? Several applications exist to compose ringing tones straight on the phone. Compose serious music? Many advanced phone models allow MIDI (Musical Instrument Digital Interface) connectivity - meaning professional musicians can use a phone to play synthesizer sounds, from the drums to guitar to piano to the violin, and you cannot tell the difference to any other professional MIDI synthesizer say from a Korg, Moog or Yamaha.

From music sound to music video

The young people of today do fully believe in the slogan "I want my MTV." Imagine now the iPod generation - for whom MTV already was the preferred

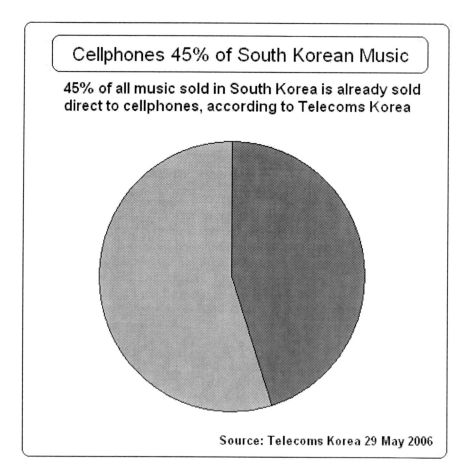

choice over radio. Yes they love the music with them. But now given the option of having only listening to music, or listening and seeing music, of course the music video wins out. Even more than at home with MTV on cable TV. Now music videos can be always with the listener. Music videos on cellphones are the inevitable future for music.

Again early numbers bear this out. In the UK, the 3G wireless carrier, "Three" of the Hong Kong based Hutchison group, has attracted about 4 million subscribers to its 3G in its first three years, out of the UK total cellphone population of 60 million. Three/Hutchison launched music video downloads to its 3G phones, which were priced at 1.50 UKP (about 2.50 USD) and in its first year sold 15 million music videos.

If given the option of buying the digital song, or buying the same song and its video, most of the MTV generation will buy the video, and clearly are willing to pay a little bit more for it. Meanwhile in South Korea, music video has been a staple of the music market for the youth on cellphones.

Do it yourself music video

And what is the end-state? If we merge videocamera-phones with the passions of bloggers and citizen journalism, and give digital tools, why not shoot your own music video? Seattle based rock band, Presidents of the United States of America have done that already in 2005. They shot the music video for their hit, "Some Postman" using only the simple videocams on their SonyEricsson cameraphones.

The issue then becomes one of content rights and mindset. The old media mindset is one of controlling media, and thus maintaining a high price. As it is exclusive, we can charge more. In the internet and mobile phone economics, this is extremely shortsighted, as Ajit Jaokar and Tony Fish argue in their book *Mobile Web 2.0*. They point out that this is a fundamental difference in old media thinking and the web experience when user-generated content is involved, writing:

> *Indeed, we believe that the requirements of the media/content industry are in contradiction to the 'network effect' application. In the former (media industry), you must restrict the free flow of content in order to make it more valuable. In the latter (applications benefiting from the network effect), you must actively encourage the free flow of 'user generated content'. Note that this is not an argument for the 'Napster mindset'. We are not advocating swapping*

> *of 'Hollywood' content but rather seek to encourage the free flow of 'user generated content'.*
> Ajit Jaokar & Tony Fish, *Mobile Web 2.0*, 2006

That phenomenon of user-generated content is already in full swing, influencing the success of music artists and soon all media. Today hundreds of bands have already used cameraphones to capture video for their music videos and often then upload these to video sharing sites like YouTube. The British dance label Ministry of Sound actually invites fans to submit videos for their 2006 hit ("Put your hands up for Detroit" by Fedde Le Grand).

D ENJOY LIVE MUSIC

Again the extent of how far music can be enjoyed on the phone is not limited to replicating what is possible on CD or DVD, You can also view live

---

4% of Americans listened to MP3 songs on their cellphones in 2005

Source: TNS November 2005

---

concerts on 3G phones - and remember in South Korea already more than half of all phones were 3G by 2006. Robbie Williams was the first major artist to multiply hundredfold the audience of his premiere rock concert broadcast live in Berlin at the launch of his newest album. His concert was simulcast live on 3G phones into a number of countries.

Learn to dance

Dance is a significant part of enjoying music. Yes you can dance while wearing your stand-alone MP3 player or iPod. But that MP3 player cannot teach you to dance. However, since it was launched in South Korea, the virtual dance tutor is on the phone in many countries. Select your favorite song, set the tempo, and a stick figure will guide you through the steps. As you learn your dance moves, you speed up the tempo. The virtual dance tutor has been on South Korean cellphones for three years already. Rihanna the

American hiphop artist was one of the first western artists to adopt this idea in 2006. Moreover, if you think of yourself as a choreographer, you can capture your dance moves and share your skills with others with a 3G phone. Again thousands of clips on YouTube validate this concept.

The British pop trio Sugababes was among the first Western bands to invite its fans to submit video of dance moves that the Sugababes would then perform in their stage act during their international tour of 2006. Meanwhile we remind readers that Cyworld in South Korea actually attracts more video uploads than YouTube even though South Korea has only one-sixth the population of the United States. It is all due to the penetration of the 3G cameraphones.

E CYWORLD AND MUSIC

We have discussed Cyworld in much detail in chapter about virtual reality

> ## 26% of South Koreans listened to MP3 songs on their cellphones in 2005
> Source TNS November 2005

and mention it in many parts of the book. But a brief mention must be made in the music chapter also about Cyworld. The virtual playground of Cyworld is a natural environment for the consumption of music. If you remember, a major element of the Cyworld social networking experience is the Miniroom for each user where they can interact with friends. Being a good host means you make your Miniroom pleasant. One of the best ways to do that is by the music you'll play to your friends.

Background tunes

This has turned into a bonanza for Cyworld. By 2005 background music for Cyworld was generating sales of 100,000 songs per day. At about 43 cents per song, the ambience nature of Cyworld produces music revenues of 18 million dollars annually, shared between Cyworld (i.e. SK Communications) and the rights holders in the recording industry.

Welcoming tunes

The latest variant of the music in Cyworld is yet another variant of the ringing tone concept, called the Welcoming Tune. The owner of a Miniroom in Cyworld can designate a given song to be played whenever someone enters the room. It may take a little while for the owner to come to the room to keep company to the visitor, so the welcoming song is a good sign you are a good host, are on your way, and give something nice to your guest just upon entry. Waiting Tunes are full-track MP3 files that are played once, and each time they are played, they also generate a charge (40 cents per play). This is like the modern re-incarnation of the jukebox. Every time you play the song, a charge appears on your phone account. Another big hit inside Cyworld and provides a new way to enjoy music digitally.

Big ecosystem for digital music

While we have discussed some handsets and the cellular networks so far, we have to point out that the South Korean digital music industry is a big ecosystem with dozens of companies all collaborating to build this new industry. They include big players like Soundbuzz already active in 13 Asian countries, Cowon with its JetAudio music platform, and Reigncom which makes the iRiver MP3 player. But also many other companies are innovating in the music space, such as Navermusic, 52street, Beatbox, IlikePop, Bugs, Dosirak, Funcake, Jukeon, MaxMP3, Mukebox, Musicon, Muz, Mylisten, OIMusic, Tubemusic, SM Entertainment and Wavaa.

F FIGHTING PIRACY

Just like Napster, KaZaA and other file-sharing systems that spawned illegal file sharing worldwide, South Korea also had its share of file sharing services. But also, understanding digital rights and the role of content ownership in the future of the digital entertainment world, South Korea moved rapidly to remove these, what were seen as parasites of the new digital music industry. After the first battle against the biggest file-sharing service in South Korea, Bugs Music was won, the last of the big players in that space, Soribada the peer-to-peer online music site was shut down in November 2005. Meanwhile Bugs Music has turned into legitimate music sales and is now working with the cellular network carrier LG Telecom.

In addition, other digital trends appear in the music field. For example the popularity of search and social networking (digital communities) are obvious in the digital music space. In South Korea for example the search engine Daum has introduced many features that help find music and works closely with the music portal site Muz.

Export of Korean Music Artists

The so-called "South Korean Wave" has swept over Asian countries in recent years perhaps best known for the popularity of South Korean soap operas. As part of this wave many South Korean music artists have found massive fan bases beyond the Korean shores and millions of people now listen to South Korean pop stars. SM Entertainment is the music label behind some of the hottest South Korean music acts including BoA, Shinhwa and HOT who are all huge for example in Japan.

Current developments

The music industry is experiencing innovation in all parts of the world, from the USA (iTunes) to Finland (ringing tones) but more than half of all the digital innovations have come out of South Korea. Because of the high penetration of broadband and 3G cellphones, as well as a vibrant domestic music industry suddenly discovering a big demand abroad for its artists, South Korea has an optimal incubation lab for new innovation for digital music.

Pearl - Karaoke on cellphone. So you want to sing the song, but can't remember the lyrics. South Korean wireless carriers all offer karaoke services with vast libraries of all the karaoke classics.

Some of the developments currently explored in South Korea include sales and delivery of music over the DMB broadcast technologies and/or with the networks, digital radio over broadband (similar to IPTV on the television side), various multimedia adaptations to cellphones in for example playing music you selected, on the phone of the person you are calling.

An important note is to remind readers that the music industry needs Digital Rights Management (DRM) solutions. Here too South Korea is taking a very active role with INKA Entworks deploying DRM solutions such as their Netsync. The DRM solutions are very advanced, not limited to only one player platform, such as that on the Apple iPod, and DRM is interoperable with MP3 players, musicphones, content providers, network operators and conforming to international standards.

The music opportunities in Digital Korea are enormous. To understand the importance of music, 85% of South Koreans own an MP3 player (the category includes iPods), but out of those, already 57% of South Koreans consider the musicphone as their primary device for the consumption of music.

Case Study 8
Melon Music

While iTunes is the best-known online music service in the world Melon in South Korea is arguably the most advanced music service in the world., Melon, provided by the South Korean digital services innovator WiderThan, was the first fully convergent music portal serving both broadband internet users and music-playing 3G cellphones. Melon features over-the-air music downloads, real-time music listening, and the whole service is available on cellphones, personal computers and stand-alone MP3 players.

We need to point out that on high speed 3G networks - and now the new 3.5G (HSDPA) networks, the download speeds to acquire music are short. Music purchases are often impulse buys. A song is heard on the radio, seen on TV, heard at a party or club, and an urge to buy the song is formed. Then the faster the song can be delivered, the more likely it is that the song will be bought. This is what the benefit is to provide music direct to cellphone owners. The immediacy of the purchase, when the impulse arrives.

Being part of the biggest South Korean wireless carrier, SK Telecom, Melon Music also reaches the largest share of the South Korean music consuming customer base. Recently Melon announced a partnership with Intel's Vive digital home platform to give its users access to TV content as well.

As Melon had 4.2 million members by the summer of 2006, out of SK Telecoms subscriber base of 19 million, this means 21% of its subscribers have signed up for Melon Music. Or to put it in another way, out of SK Telecoms world-leading proportion of its subscribers on 3G, four out of every ten 3G subscribers have taken up the Melon Music service.

Melon Music offers MP3 files for purpose, of course, but Melon Music also has a flat rate option for consuming unlimited music and paying only a monthly fee. Melon had 4.2 million members in 2006 who were buying music. Out of the 4.2 million about 600,000 were on the subscription plan.

Melon music earned 45 million dollars in 2005 and was on target for 70 million dollars in revenues from its various music services for the year of 2006.

Chapter X
Pervasive Computing

Ubiquitous Connectedness

Image courtesy *IT Korea Journal*

> "There are three kinds of death in this world. There's heart death, there's brain death, and there's being off the network."
>
> **Guy Almes**

X
Pervasive Computing
Ubiquitous Connectedness

The most connected country. The internet country with highest speed. The wireless country that is already moving past 3G and WiFi. In the most concrete ways the most digital country in the world. How can we describe South Korea? The locals have adopted the term ubiquitous, to describe how digitalization has reached everywhere, how computing is ever-present and connectedness is omnipresent. It is the most advanced digital country in the world, what seems like stepping into the future to visitors from all other countries. This chapter examines where and how the internet, computing and digital systems exist in South Korea. We start with Broadband internet.

A MOST BROADBAND

The statistics are out of date almost by the time they are published. Therefore, it may be more relevant to pick a moment in time and contrast South Korea to its nearest rivals, and perhaps a few of the biggest economies.

At the end of 2005, according to the ITU and counting all forms of broadband access, South Korea had passed 50% penetration for broadband, per capita (not per household). Hong Kong and Japan were the nearest rivals at 33% and 32% respectively. South Korea had a 50% lead over its nearest rivals! Highest European countries by broadband penetration were Italy and Sweden at 29% and 28%. Canada was at 20% and the USA at 18%.

Penetration

Another way to look at the impact of broadband is to contrast it to internet penetration (i.e. versus narrowband or "dial-up" internet use). In other words to measure the migration of internet users from narrowband to broadband. Again South Korea leads. According to the ITU data, by 2005 South Korea had become the first country to achieve full 100% migration of all internet users to broadband. The tiny European country of Estonia was second at 96%, with Belgium at 81% and Israel at 80%. All are countries with less than a fifth of South Korea's population. Of major economies Canada was highest with 75% migration and Japan with 65%. South Korea's lead in broadband internet use is enormous.

Already in 2005 most households had access to two or more technologies to deliver broadband. ADSL broadband is available to 90% of the homes and cable TV based broadband reached over 60% of homes. Most South Koreans live in urban areas in apartment houses. In addition, many of these offer Ethernet based apartment LAN (Local Area Network) based broadband. Numerous wireless providers offer broadband on WiFi, WiBro and 3G technologies.

Highest speeds

A survey by Analysys of the speeds and costs of broadband in October of 2006, found that South Korean broadband top speed were the highest, average sustained speeds second highest, and the prices lowest in the world. Is it any wonder that multiplayer gaming, virtual worlds, e-government and the digital home thrive in this environment?

At the end of 2005 VDSL speeds of 20-40 Mbit/s were available to many South Koreans at just under US$ 50 a month with average speeds in the country at 4 Mbit/s. The government was aiming to reach 20 Mbit/s to all homes by end of 2006 and this target seems to be reached (official data was not available when we went to print). This speed benchmark is relevant as 20 Mbit/s is the speed needed for viewing HDTV on broadband. The country of "bballi bballi" (hurry hurry) is so addicted to higher speeds it insists HDTV broadcast TV speeds when most countries do not even offer HDTV programming yet. Moreover, this is not the ultimate speed. "High speed broadband" already sold in South Korea in 2006 was 50 Mbit/s and first

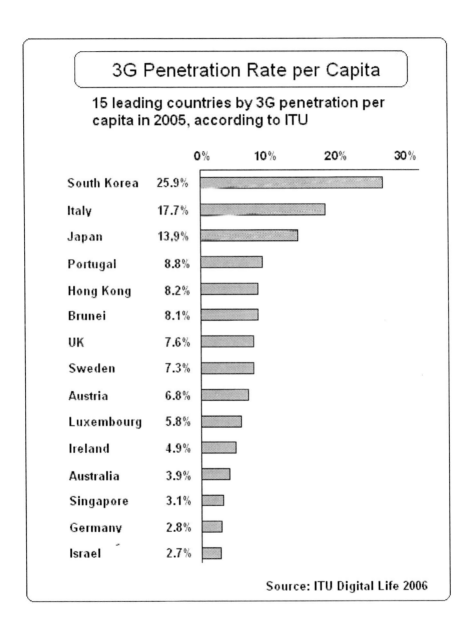

offers appearing for 100 Mbit/s, which should be the norm for all homes by 2010.

While it is fifty times faster than the common definition of broadband speeds, even 100 Mbit/s is not the ultimate top speed. South Korea has started to experiment with gigabit speed broadband with the first sixteen 100 Mbit/s broadband connections in trial operation. This leap into future broadband speeds will be our last case study in this book, at the end of the digital convergence chapter.

B 3G CELLPHONES

Another measure of the state of advanced technology is the extent of advanced 3G networks. NTT DoCoMo in Japan launched the world's first 3G network in October 2001. Only three months later South Korean SK Telecom was the second carrier to launch 3G, and by May 2002 South Korea became the first country with competition in 3G as KTF entered the 3G market in South Korea. Today South Korea leads the world in migration of subscribers

---

8% of British cellphones were 3G in June 2006
Source: Informa 2006

---

to 3G, with over 60% of all subscribers migrated to 3G by end of operating both international standards. By comparison, European leaders Italy and Britain, which launched 3G in 2003, have had fewer than 20% of their subscribers migrated to 3G by end of 2006.

2D Barcodes

We have discussed the migration of internet access from access from personal computers to access from advanced 3G cellphones, and already today in South Korea the majority of internet access is from cellphones. Here a critical enabling technology has been the 2D barcode. We think the 2D barcode reader is the single biggest change agent in how cellphones are used to access web content. Iconlab Chief Operating Officer, KW Park explained in interviews to this book on the relevance of 2D barcodes:

> *"The 3G mobile phone is a very convenient device but it is difficult to use phones for internet access because of the limits of the keypad. Now with the 2D barcodes like we see already in South Korea, web access becomes easy."*
> KW Park, COO Iconlab

We discussed the 2D Barcode in depth in the case study in the shopping chapter earlier in this book. For this Pervasive Computing chapter, we want to stress than 2D Barcodes allow immediate internet access from 3G cellphones without ever having to "triple-tap" long web addresses on the cellphone keypad. This is a key factor in enabling ubiquitous computing on the cellphone and helping usher in the age of the wireless internet on cellphones.

HOW ACHIEVED?

39% of South Korean cellphones 3G by June 2006
Source Informa 2006

There are many reasons why South Korea has achieved this success. One clear cause has been the determined leadership by the government. South Korea has wanted to achieve leadership in the digital IT industry and the government has focused effort and resources to this goal. A vital element has been collaboration. South Korea has a history of companies working in clusters around a given industry leader or technology. In a similar fashion, whole sectors of the South Korean telecoms, computing, broadcasting, gaming, programming, music, etc industries have come together to collaborate and cooperate, but also to compete. The efforts of academia have been important, with various South Korean industry associations and the universities devoting resources to this goal.

Geography helps

However, there also are some other aspects. Geography plays a part. Half of South Koreans live within the metropolitan area around Seoul. The few other major cities contain most of the rest of the population. Thus creating the high speed digital infrastructure has been perhaps more similar to wiring a city-state like Singapore or Hong Kong, than that of the true mid-sized country with physical size of South Korea of 98,000 square kilometers (38,000 square miles) or a bit larger than Portugal, slightly smaller than the state of Indiana.

> **In just one year from launch, and without subsidies, 5% of South Korean cellphone owners had upgraded to digital TV tuner cellphones**
> Source: KIPA September 2006

Some of the factors may be socio-cultural. A large proportion of the South Korean population is young, but very highly educated. They would therefore be still willing to experiment (typical of the young) but also appreciate the value of modern technologies. South Koreans seem to appreciate gadgets and technology, and as a population they seem to be competitive. These would set it apart from for example Japan, also highly educated and competitive, but as a society, the population has been aging for several decades so the proportion of the young in Japan is much smaller than that in South Korea.

Collective belief

Moreover, perhaps part is a collective belief in achieving global leadership. South Koreans saw Japan do it in various technologies from cameras and wristwatches to home electronics, automobiles and robotics. They had an example very near. However, with the World Cup of Soccer and the recent rapid global success of brands such as Hyundai, Samsung, LG and Daewoo have given confidence that South Korean industry is globally competitive. Then the recent South Korean export of popular culture, from soap operas to pop stars has added fuel to this belief. South Koreans are very proud of their accomplishments so far.

Now with again the government's strong support, the current, connected generation wants to make an even larger impact on the world stage. They know the future belongs to digital leadership. That is what it seems the whole society is single-mindedly pursuing. As we said before, one of the most used services on mobile internet is self-improvement, learning English, programming skills, web design etc. There seems a broad consensus among South Koreans that the whole society will benefit, as the country becomes Digital Korea.

Virtuous cycle

Finally there is a virtuous cycle of working in the technologies that enable digital infrastructures. By developing broadband, wireless internet, 3G, DMB and robotics technologies and their various elements, the companies that develop these will also be among the first to understand how to gain from them. It is like Nokia's famed "PowerPoint Palace" (the Headquarters of Nokia) in Espoo Finland. Being in the most "mobile" country i.e. highest cellphone penetrations over the decades, the HQ of Nokia is naturally also the most advanced office for wireless telecoms services. The suppliers of telecoms services in Finland have most demanding customers but Nokia is the jewel of any Finnish company's reference customer list. Therefore Nokia is bound to get the ultimate in mobile technology for its own use. This both helps its own engineers and benefits the suppliers by having a most demanding customer help them develop the next evolutions. The same would be true for internet services at Cisco HQ, etc.

Thus the South Korean companies gain from the most connected and digitally enlightened work force. The South Korean consumers gain from being the home market for several of the world's leading suppliers in the digital environment. The South Korean industry can sell ever more advanced services and products, as their local customers are knowledgeable enough and demanding enough to want the cutting edge. A virtuous cycle promoting early use and rapid adoption, as well as user-driven product development.

D IDEAL TEST BED

This makes Digital Korea also an ideal test bed. Test beds are crucial in the high tech industry for maintaining a competitive edge. Typically most innovations take time to mature. The first version of most technologies is not yet optimal, and experimentation in the market with early adopter customers is needed to weed out the problems and perfect the product. Pioneering companies often do not make much money for their inventors and find only a small customer base. It is only the second, third and subsequent generations that reach mass markets and make money. To bridge the gap between initial invention and customer appeal takes trial and error. For that a good test bed is critical.

Any company internationally with the right technical product but without the audience or infrastructure for testing can find optimal conditions in South Korea. First of all the technical platforms already deployed are literally the world's best, starting with DMB, WiBro, CDMA 1x EVDO, WCDMA and HSDPA. There is a well-trained work force knowledgeable in installing and maintaining these systems and their services and support technologies.

> **One third of South Korean student population send 100 SMS text messages every day.**
> Source: *Korea Times* 9 Feb 2006

As a test bed South Korea also offers other valuable benefits. Again the central role of Seoul helps in distribution. Half of the population lives within an hour's drive of the center of Seoul. With Incheon the port city supporting the metro area of Seoul, shipping goods to and from Seoul is easy. And the language? This is also a very good point for any test bed. Korean is not spoken in any other country except North Korea. So should initial tests prove disastrous, these will be discussed mostly only in a language nobody overseas can understand.

Western companies coming in

Numerous Western technology companies have already discovered South Korea and set up research and cooperative initiatives there. A good example is Canadian telecoms infrastructure vendor Nortel that has set up a joint venture with LG of South Korea. Nortel's Peter MacKinnon, Chairman LG-Nortel JV, explained some of Nortel's reasons in an interview for our book:

> *"We wanted to be more involved in South Korea as a market because its customers are so demanding and the market is so responsive and competitive. With the wireless carriers reporting world-leading data usage levels and adopting services that are not even requested by customers in other markets, we felt South Korea*

was the future for the telecoms business, and wanted to be able to experience it directly. The joint venture with LG gives us the best opportunity to bring in our experiences while learning from a company well familiar with the market."
  Peter MacKinnon, Chairman LG-Nortel JV

South Korea offers a rich opportunity for Western companies as it features a highly educated population, where English is widely understood and spoken. Digital Korea adds the experimenting and curious nature of the South Koreans when faced with new technologies and innovations. In many ways South Korea is one of the most attractive locations to do technology market experimentation.

E PRIVACY

With the ubiquitous nature of the digital world, South Korea is also first to witness abuses and the downside of the innovations. Criminals are fast to adapt to cyberspace. With more digital devices connected to each other, matters such as privacy and security become ever more important. South Korean suppliers are now implementing security means such as fingerprint scanners and voice recognition to laptops, PDAs and cellphones. As the cellphone becomes the focal point for identity, commerce and even holds our digital keys, the secure nature of all this access to our personality and property is ever more relevant. Yes, biometric technologies add to the cost of our digital access devices, but that is a reasonable price to pay when so much is guarded behind our pocketable devices.

Selling sauna snaps

With any new technology come abuses. One South Korean experience is the mok yok tang sauna baths. Men and women bathe separately and move from sauna to pools of hot, medium and cool water. A recent scandal involved a woman secretly taking pictures of her fellow nude bathers with her cameraphone and then selling the nude snaps to a website. In the resulting furor the MIC introduced regulations requiring a loud sound that cannot be muted when pictures are taken, to all cameraphones sold in South Korea.

Role of communities

As information technology becomes pervasive, ubiquitous, it also promotes the collaboration between people using the technologies. In that kind of environment, as humans are still social animals, we use the power of the new communication tools and methods to form groups, digital communities. The rise of digital communities and how that impacted all aspects of society was discussed by Ahonen & Moore in their book *Communities Dominate Brands*. In the book they explain that their very connectedness makes digital communities powerful not only in forming, but also at automatically improving themselves:

> *A key element of virtual communities is that they can immediately implement what they think of. Any innovation within a community can be decided upon by the community and then implemented. Very importantly for businesses and brands that sit at the other end of the power struggle, communities are not only fast at mutation from ideas within, they are also remarkably fast at learning from the outside. As communities are connected - by some number of degrees of separation - to every other community, soon a good idea in one country is copied in another. The contagion effect is remarkably fast.*
> Ahonen & Moore, *Communities Dominate Brands*, 2005

Communities in a digitally ubiquitous society will play an ever-increasing role. Many Western media sources discussed various surprising phenomena from bloggers forcing CBS news anchorman Dan Rather to resign, onto companies setting up corporate islands in Second Life. We see these events already in South Korea in the powerful influence of Cyworld, Ohmy News, Lineage etc.

The end of email?

South Korea has been one of the first countries to observe that the youth and young adults prefer SMS and IM to email. Many have commented on this, and similar findings have been reported from Europe to the USA. However, while most countries report a slight decline in email usage over the faster communication of SMS and IM, South Korea is now reporting an almost complete abandoning of the use of email by the youth and young adults.

The first survey by Lee Ok-Hwa of Chungbuk University of 2,000 middle, high school and university students found that more than two thirds replied "I rarely use or don't use email at all". It is again the urgency, "bballi, bballi" (hurry, hurry). Email is too slow, too formal. In addition, the recipient is not necessarily connected.

SMS Text Messages Sent

Comparison of SMS Text Messages sent by cell-phone owners in USA, UK and South Korea in 2005

Country	SMS sent per month	SMS sent per day
USA	18.5	0.6
UK	39.7	1.3
South Korea	268.7	8.9

Sources: MDA, CTIA & NIDA

Only through SMS can you reach anyone immediately. Moreover, if you are both connected, then of course the youth communicate using IM rather than email. As we reported in the Youth chapter at the beginning of this book, 30% of the youth, the heavy users, average 100 text messages sent per day.

F CELLPHONE THE REMOTE CONTROL

Samsung has been developing cellphones for various remote control functions. It makes sense as in South Korea Samsung is one of the major appliance manufacturers such as refrigerators and air conditioners etc. Each home appliance can be equipped with a network card and internet address. Then systems such as a home gateway for example using the electricity power lines already existing in the wiring for the home can transfer the signals to the individual appliance. The home gateway can itself be controlled by the cellphone. We discussed some of these technologies earlier in the book in the Digital Home chapter.

Cellphone rules of conduct

As cellphones have become ubiquitous, and they do cause changes in how we behave, from speaking suddenly as if to ourselves, to possibly disturbing others with the loud sounds of our ringing tones, the phone has created cause for reconsidering what is polite and what is not. To address these issues, KTF the second largest wireless carrier of South Korea has released its eight rules of conduct with cellphones. These seem particularly well thought out and we can warmly endorse them:

Cellphone code of conduct:
8 simple rules from KTF in South Korea

1. Switch to silent mode in public places.

2. Make your conversation quiet and simple.

3. Switch to silent mode in a class or meeting room

4. Ask the person you call if he or she can answer the phone at the moment before you start conversation

5. Don't start conversation with someone who is driving. Call back later.

6. Refrain from using mobile devices around medical equipment or during a flight

7. Identify yourself when you send a text message

8. Care about others' rights and privacy before you use camera phone. Never take pictures of others without their consent.

Clearly these recommendations have been born in a country that has experienced issues with advanced cellphones for some time, and are a very good set for any considerate person to follow.

Internet cafes

South Korea has 20,000 internet cafes called PC bangs where the consumer can rent modern PCs with high-speed internet access for about one dollar per hour. These are also popular places to play videogames. Often a team of gamers will go to a PC bang to play online inside the same game, such as Lineage. They sit at neighboring PCs and play together in the same part of the MMOG environment. This kind of playing together is not possible at the home even with broadband, as typically most homes do not have several high power PCs next to each other.

Lessons for the rest of the world

In South Korea we have the most connected, highest speed digital broadband internet wireless society. Its ability to adapt to and take advantage of new technologies is unparalleled. South Koreans are also eager to invite companies from other countries to share in its riches and to learn from the other innovators and inventors.

> **Pearl - translate dog barking into human speech, vai cellphone.** Ever wanted to know what your dog means when it is barking? The dog-to-human translation service "Bowlingual" is available on cellphones, in South Korea. Now you can have your dog barks translated into human language in the form of SMS text messages. I'm hungry. I'm hungry.

One of the significant lessons from South Korea is that the digital divide or gap between expectations and access to digital benefits of urban citizens and rural citizens is becoming smaller. These days the barrier to digital and community connectivity and interaction via fixed or mobile access is much smaller and particularly youth consumers aspiration in smaller towns and remote villages are narrowing, be they in South Korea, China, Ireland, Poland the USA or Chile.

Infrastructure is becoming easier and less expensive to implement and at the same time increased demands of users for multi tasking, and multi-access points is increasing. The difference in behavior of South Korean youth vs. international youth in how they access digital content and services is also narrowing. Of course the Urban population of Seoul will have an advantage into the immediate future as they continue to enjoy advantages of innovative cutting edge (and indeed bleeding edge) technologies at greater pace than other urban cities. Nevertheless, due to the blogosphere, multinational online gaming, international chat boards and cross-cultural digital communities, the awareness factor but perhaps not the intensity of interactivity of Gen C is high also internationally. While others may envy the extent of the digital environment of South Korea, most of the early adopters in other countries will experience at least partially those same elements. Only not all of them on one place like in South Korea.

Case Study 9
Wearable computing

South Korean's are looking forward to the next generation of mobile computing devices, sometimes called "Post PCs". Two categories of Post PC products are of particular interest to the mobile information society in South Korea: portable and wearable.

Some of the obvious wearable applications that are already on the market include the bluetooth connected earpiece and usually connected microphone.

Post PCs can be a in the form factor or a PDA or tablet PC or the smartphone type of cellphone.

But portable computing can also be built into form factors of wristwatches, sun glasses and clothing. Various flexible display panels have already been introduced as well as technologies to embed circuitry into clothing. The display can be for the computer user to see what he/she is doing with the computer, or to allow displaying messages, images etc to other people, similar to the message on a t-shirt, but being digital, it can then be customized and changed on the fly.

What may set these devices apart from PC's and PDAs as we know them is the user input method.

The mobile Internet has long been constrained by a good method for inputting information. The PC world has a 101 key QWERTY keyboard and a mouse. The cellphone for a very long time only had the keypad, forcing triple-tapping of the keypad to create letters.

Recently the embedded camera and now the 2D barcode reader have helped get around some of the typing such as entering web addresses. But 2D barcodes do not solve the full problem. For wearable computing a new form of data entry is needed.

One possible area is projection keyboards where a small projector projects the image of the keyboard onto any flat surface and a sensor detects if a finger presses the key.

Another way around the data entry is building the keys for example into the sleeves of a shirt. Still other trials are now using motion sensors so that a user can wave his/her arms to create letters and risk looking like a deranged human windmill.

    Meanwhile on the side of the mouse input, Samsung has been working on a solution to this dilemma and has recently released the world's first wearable mouse.

Chapter XI
Multiplayer Gaming

Immersive Entertainment

Image courtesy of Lineage and NC Soft

"Once consumers interact with a media, they never turn passive again."
Gerhard Florin, Executive Vice President, Electronic Arts

XI
Multiplayer Gaming
Immersive entertainment

South Korea is a gaming mad country. Videogame stars are celebrities and the success of the South Korean gaming team in videogaming world championships is a major event on Korean TV and media. The various digital platforms, from broadband internet to 3G mobile telecoms make for a rich environment to develop games to various tastes. A telling statistic is that of the total South Korean population, 10% play videogames every day. Alternatively, another revealing statistic: one in four South Koreans has played the same multiplayer online game at one time or another. More than any other country, Digital Korea is also Gaming Korea.

A HOW BIG IS GAMING

Let us start with the money. The videogaming industry can be divided roughly in two; about half is the sale of the consoles like the Sony Playstation and the Microsoft Xbox and Nintendo Wii. The other half is software sales, the sale or rental of the actual gaming titles like *John Madden's Pro Football*, or *Grand Theft Auto*, or the *Tony Hawke* series of skateboarding games. The videogaming software sales are roughly divided into internet/PC games, and console games.
    The two most rapidly growing sectors of videogaming software are online games and games for cellphones. Cellphone gaming accounts for

about 14% of the global videogaming software revenues and about 11% are spent on broadband internet gaming.

Like in music, cellphones are emerging as the trial platform. *Urban Freestyle Soccer* became the first video console game in 2003 to be released as a mobile game first before the console version to follow. In addition, with the rapid migration to 3G phones, South Korea is an ideal environment to adopt mobile gaming. And they do. 37% of South Korean cellphone owners play downloadable games on their phones and 15% play mobile games every day. Of the content charged by wireless carriers, videogaming comes in second after music. But let us take a look at the gaming phenomenon in context.

Gaming generation

The biggest single influencing technology for the previous generation was television. While TV is also of interest to younger generations, TV especially how it was consumed in the past, is passive. Gaming is active; the gamer will enter the imaginary environment of the game, and take an active role in playing in it, with it, living it. The gamer takes control of how the story unfolds in the game, whereas watching TV or a movie the viewer cannot control the storyline.

Very similar to how older generations related to TV when they were young, young videogamers today will have dreams, even nightmares, based on the latest games they play. The game characters are introduced into the "regular play" just like previous generations brought favorite TV characters into their play such as playing "cowboys and indians" in the 1960s or "cops and robbers" in the 1970s etc.

Digital divide among gamers

In conjunction with mobile phone usage and the fixed Internet, gaming has created a cognitive divide between baby boomers and the Generation-C. For example Americans now spend more on video games per year than going to the movies. Put in another way three out of four American households with male above the age of eight has a video game console. Sony's Playstation games console alone has a place in 25 per cent of all US households. Games are no longer "exotic" or "interesting" – they are an established feature of the media landscape.

Uploading Pictures from Cellphones

Proportion of the population who directly upload pictures from cellphones to social networking sites like Flickr, MySpace, Bebo and Cyworld

Country	Number of People uploading Pictures	Percent of total Population
USA	12 Million	4%
UK	3 Million	5%
South Korea	21 Million	30%

Sources: Telephia TAM Report January 2007 and Seoul Magazine December 2005

Impact of mobile gaming

Where it comes to cellphones, for a long time videogaming was thought to be limited only to a pass-time element, and work mainly as simple built-in games that shipped with the phone. Typically this would be Nokia's *Snake* game or something like *Tetris*. Mobile telecoms experts and authors Ajit Jaokar and Tony Fish explained how this changed with the advent of the Nokia n-Gage, and how it altered the perception of a handheld device being practical for gaming, in their book *Open Gardens*:

> *Will users use one device for many functions or will they use different devices each specialized for a specific function? The Nokia n-Gage was a seminal device since it was the first large-scale attempt to combine a gaming device plus a phone. Other variants are possible - and iPod with a phone and so on.*
> - Ajit Jaokar & Tony Fish, *Open Gardens*, 2004

Obviously such converged devices are now appearing in all kinds of forms, and while the n-Gage series has since been discontinued by Nokia, the n-Gage capabilities are now being introduced to various N-series multimedia phones by Nokia. Similarly the n-Gage set the stage to promote the market for the Playstation Portable. Where n-Gage sold some 2 million units, the Playstation Portable has already sold 20 million units, fully validating the concept of a pocketable modern gaming platform, even if Nokia was unable to capitalize on its first attempt.

Immersive

What is more, videogaming is immersive: it requires our full attention. Differing from how we consume radio content while driving our car or the TV might bring background noise while we read a newspaper, videogaming

> **21% of American population have set up a personal profile in MySpace**
> Source: MySpace 2006

requires full attention. If we do not press the buttons, our character does not move in the game. It requires gamer activity for the game to move along. As to its intensity and interaction, videogaming is massively more appealing to young people than passive media such as TV.

B MASSIVELY MULTIPLAYER

An extreme form of immersive videogaming is the MMOG Massively Multiplayer Online Game (often also known as MMORPG Massively Multiplayer Online Role Playing Game). The first significant MMOG appeared in 1997 with *Ultima Online*, had 250,000 subscribers. By 2004 games such as *Everquest* reached half a million active players each paying 13 USD per month just as the subscription fees. *Everquest* earned 78 million dollars per year, which placed the videogame in the class of a successful Hollywood blockbuster, and suddenly the world took notice.

Today the games most often referred in the Western media are *Everquest 2*, *Final Fantasy*, *CounterStrike* and of course the biggest of the Western titles, *World of Warcraft* with its eight million users. *World of Warcraft* is also one of the most popular Western gaming titles in South Korea. However, as we already showed in our case study of the Virtual Worlds Chapter, South Korean MMOG *Lineage II* is the biggest of them all.

Massive realities

MMOG environments are unlimited with "realities" of the real world. Players typically set up characters and can have characteristics that are different from themselves - such as a male user taking on the role of a female - and depending on how the gaming environment is set up, players can have magical and superhuman powers. Inside MMOGs, depending on their own

> ## 43% of South Korean population have set up a personal profile in Cyworld
> **Source Cyworld 2006**

rules, it is possible to fly, to be invisible, to be reincarnated, etc. Regular rules of the physical world do not apply, *at least directly*. Nevertheless, MMOGs have their own rules. A certain type of monster can kill you, etc. In fact much of the MMOG world capitalizes on the virtual world's escape from reality. Like board games such as *Dungeons and Dragons* in the past, MMOGs offer the chance to game-play with magic spells and supernatural powers against demons and monsters, facing knights and saving the princess or discovering treasures, etc.

A game world can set the boundaries of what can be done, i.e. what it costs in a given game's money currency to buy a given good or service, such as buying weapons, a home, some furniture, or to book passage from one part of the MMOG world to another, etc. The virtual world collides with the real world when there are enough players to try to "cheat" and in MMOGs there invariably are so many players that real rules of the economy come into play.

Earn goods or purchase goods

Currently on major MMOGs many players may want to purchase some goods or services using real money rather than accumulating gaming money. A gamer might run out of ammunition in the MMOG. The gamer could then go onto eBay and pay real dollars to buy more ammunition. This is like playing the traditional board game *Monopoly* with your children and finding you are losing. Then you take a bathroom break, sneak over to your neighbor, buy his game to get more *Monopoly* money and return to your game against your kids but being rich in the game again. In other words you would be paying real cash to gain more *Monopoly* money.

Yes, that certainly sound like behavior of the seriously deranged if you were just enjoying a few hours of family entertainment with a board game. But on MMOGs that really does happen on a daily basis. You see, it is not just a few idle hours of entertainment that is at stake. The players may have spent weeks or months developing their gaming experiences, building their properties, growing the abilities of their characters etc. This makes the time invested quite significant and motivates players to shift real world resources into the game.

How big?

Bizarre? Perhaps to us of an older generation, but for the gaming youth familiar with virtual worlds, this is "rational" behavior. How serious is it? Just on eBay in 2003 there were 28,000 trades every week on virtual properties for just one MMOG, *Everquest*. A completely new profession has emerged around this, called farmers of videogame gold. We discuss it later in this chapter in the section on professional gamers.

But consider it in this context. Lets assume you trial *World of Warcraft*. You find it addictive and play it for several weeks, progressing to the 25th level. Then for some reason you decide to quit playing. Perhaps your family life is suffering or whatever reason. If you know there are willing buyers on eBay ready to pay you 50 dollars for your level 25 character, to continue the quest (*World of Warcraft* goes to 60 levels), isn't it only reasonable behavior to sell your character than simply throw all that work away?

eBay transactions for MMOG goods and services involve the exchange of typical virtual properties such as weapons, ammunition, real estate properties such as mansions and castles, as well as more obscure gaming "valuables" such as magic spells. Most relevantly the gaming currency can be bought and sold on eBay. The trade in virtual goods for

MMOGs had exploded to 880 million dollars by 2004 according to *Wired Magazine*. We are on the verge of a billion dollar industry where all of the value is virtual, and the value was completely created by the gaming community in their spare time.

Measured in real dollars

Better yet, with the sales of virtual money on eBay, an exchange rate has emerged for what the gaming currency is worth in real money. The exchange

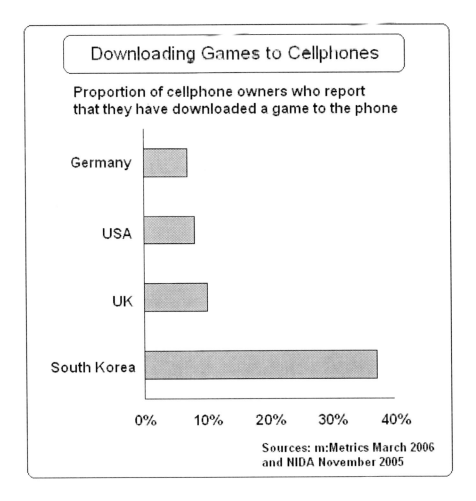

rate for the "Norrathian Platinum", the gaming currency in *Everquest*, was 1 538 to a dollar at the end of December 2003. Now properties in the game can be valued in dollar terms and real values established for virtual goods. These are determined by laws of supply and demand and reflect economic realities in the gaming environments. *The Financial Times* reported that the virtual GDP of *Everquest* could be calculated per (virtual) capita and at 2,266 dollars per capita, the virtual gaming world has created wealth that made it the 77th wealthiest country on Earth.

We want to stress this point. In just one MMOG virtual world that had half a million subscribers, three years ago, its gamers had ***created*** an economy so wealthy, it produces the annual equivalent to the ***total domestic product*** of a vibrant country the size of Croatia, Cuba, Ecuador, Liberia, Slovakia, Sri Lanka, Syria, Tunisia, Uruguay or Vietnam. Imagine how much more can be created in more modern worlds that have many millions of

---

By 2006 *World of Warcraft* had 8 million subscribers worldwide

Source: World of Warcraft

---

gamers today. The virtual world opportunity is immense, and growing at unprecedented speeds because their economies are not limited by the limits of the physical world. Many real world start-up companies exist already to provide goods and services within that virtual world and the opportunities are there for mainstream business to get involved and active. The virtual environment is the new "gold rush" of this decade. Anyone can join in.

C ENTER THE COLOSSUS: LINEAGE

Today the best-known MMOG in the West is World of Warcraft with eight million users and frequent coverage in the world press, but the biggest MMOG in the world however, is *Lineage II* from South Korea with 14 million customers. The game was developed by NC Soft. The first version was launched in 1998 and *Lineage II* was released in 2003.

A typical role-playing multiplayer environment set in the Middle Ages, *Lineage* is based on a comic book resembles previous games such as

Dungeons and Dragons and the gamers can appear as wizards, elves, knights, princes, etc. Today Lineage has gaming sites active in South Korea, China, Japan, Hong Kong, Taiwan, UK, Germany, France, Brazil and the USA. Obviously almost twice as big by users as its nearest rival, *Lineage* is the colossus of the videogaming world. Everything discussed in the above about MMOGs exists in *Lineage*, only at an even more massive worldwide scale. We discussed *Lineage II* in the Case Study for the Virtual Worlds Chapter, so we will not repeat that part here. Nevertheless, everything that applies to MMOGs that we wrote here in this chapter, applies also to *Lineage*.

D CASUAL MULTIPLAYER

Another multiplayer gaming opportunity is in *Kart Rider*, which we will discuss in the case study to this chapter. Briefly, *Kart Rider* is a driving

> **By 2006 *Lineage* had 14 million subscribers worldwide**
> Source Lineage

game. It has taken South Korea by storm. With over 12 million South Koreans already having driven a car in *Kart Rider*, this game set the world record for highest national penetration for a single game. *Kart Rider* has been played by 25% of the total South Korean population! Yes, *Lineage* has more users at 14 million, but they are spread across a dozen countries. *Lineage* is played by about 6 million South Koreans.

Kart Rider is not played for immersive sessions of hours on end, but rather as casual gaming, short periods, a quick ride, a quick race. In addition, very importantly a large proportion of its gamers have not played online games before, so it *Kart Rider* helps bring new gamers into the multiplayer environment. However, we will discuss *Kart Rider* in more detail in the Case Study.

E PROFESSIONAL GAMERS

Videogaming is becoming ever more serious. Today world championship tournaments exist in several of the most played games such as *CounterStrike, Quake4, World of Warcraft 3* and *Project Gotham Racing 3* and *Need for Speed: Most Wanted*. With professional gamers from literally around the world so too are the various winners. Still, South Korean gamers are taking a major haul of the trophies.

World champions

During the 2006 gaming season in the *Warcraft 3* category for example Jung Hee "Sweet" Chun won one gold and one silver personal medal as well as one gold, one silver and one bronze team medals. Jae Wook "Lucifer" Noh won one gold, one bronze personal medal, and three gold team medals; Jang "Moon" Jae Ho won one personal gold and three team gold medals. Other star South Korean medallists in world tournaments in 2006 include Dae Hui "FoV" Cho and Daeho "Showtime" Kim. The South Korean team won three of four team gold medals and brought in the bronze in the fourth team tournament.

> **Pearl - Virtual girl friend on cellphone.** The virtual girlfriend/boyfriend is not used mostly by adult single techie nerds, but rather by young teenagers who haven't yet had a real relationship. They use the virtual partner to learn how to be a good boyfriend or girlfriend.

These gamers are not playing only for the fame of winning gold in e-sports tournaments. They earn salaries from sponsorship deals with incomes of 100,000 dollars per year and above. Some of the tournaments feature prize purses where the winner may win that amount at a single tournament. Digital Korea? Gaming Korea! They do take their gaming very seriously.

Farming for Videogame Gold

However, mastering a game and attracting sponsorship money is not the only way to earn a living in gaming. Many average skill gamers turn professional by becoming farmers, farming for videogame treasures, also known as farming for videogame gold. This tendency was first discovered as a pattern among users in *Lineage*. Farmers based in South Korea and increasingly in less wealthy parts of China started to gather the various "consumables" of videogame play and then go to outside auction sites and sell the goods.

> ***Kart Rider* is so popular in South Korea that it has two cable TV channels dedicated to following the online broadband multiplayer game**
> Source: Nexus

Therefore, in practical terms, there are several valuable commodities in the game to help the gamers along in the game. Treasures. Some are weapons and ammunition, some are sustenance such as food or medicine, and others are gaming gimmicks such as spells or the opportunity to become invisible. Typically there are certain locations in the gaming world, where this type of treasure is hidden. When the gamer arrives to the location, and at times performs some ritual such as touching a panel or entering a key etc, then the gamer gains that treasure.

Cheat the gaming logic

The gaming logic is usually so, that a gamer can only get a certain amount of a given treasure at any one time. Then there is a timer that resets that treasure and it "regenerates" so another such treasure is available.

A gamer who carefully monitors such treasure sites learns how rapidly those treasures regenerate and soon learn how many minutes or hours it takes before another treasure can be collected. Alternatively, the gamer discovers other locations where more treasures can be acquired. Then the gamer can collect enough of the treasures and go outside of the game, to eBay or another trading website, and offer these treasures for sale. Exchange the dollars and one gamer has money, the other gamer has new Armour, ammunition, spells and food, or whatever was being traded.

Gold Farmers can earn a good white-collar employee level salary by farming the treasures of the games full-time every day. Thousands of people make money specializing in most major MMOGs already. The game developers try to limit this, but it is a continuous game of cat and mouse, with the money being too strong an incentive, the farmers mostly winning. Sometimes one is caught, but within minutes that person is back on a new identity, perhaps from accessing from the neighboring internet cafe, and the farming continues.

Valued at 830 Million Dollars in Korea alone

The various money streams from Farming and other money transfers between the virtual worlds and real money, inside South Korea alone, in 2006, were estimated to be worth 830 million dollars by the Korean Game Development and Promotion Institute. Only two years earlier the total value

---

Value of virtual property sold in South Korean MMOGs for real cash in 2006: $830 million
Source: Korea Game Development and Promotion Institute

---

of all global funds spent on online gaming was about 800 million dollars. We are witnessing the birth of a legitimate new economy in every sense of the word, however virtual the actual labor inside those worlds may be.

End of the game

We have looked at avatars and virtual worlds in the Virtual Worlds chapter, and now at multiplayer videogaming in this gaming chapter. These two concepts are closely linked and have many similarities. However, it is clear that videogaming is an immersive media, and multiplayer gaming is even more addictive than gaming against a console or computer. What we do see out of South Korea is that when gaming reaches critical mass, it soon develops professional elements from the sponsored gamers to the farmers of gaming gold. Moreover, most of all, the multiplayer gaming environments, themselves not limited by the rules of the physical world, can evolve and grow dramatically.

The first *Lineage* had 4 million gamers, similar to say Ireland's or Singapore's population. *Lineage* II had already 14 million, approaching in size to populations of Chile and the Netherlands. Soon we will have MMOGs with 25-30 million gamers or more. Then we are looking at populations in the scale of Malaysia and Canada. In addition, the gamers inside these MMOG worlds are actively building honest real economic value out of their gaming activity. Where will these trends take us over the next decade? Keep your eyes on Digital Korea for the early signs.

Case Study 10
Kart Rider

Drive fast. You don't need a driver's license. You don't need to own a car. All you need is to log onto *Kart Rider* and you can experience the thrill of driving fast. An online multiplayer videogame that has become a national phenomenon, it now has its own TV channel and top drivers have sponsorship deals.

Kart Rider is the biggest hit game in South Korea. 12 million people, a massive 25% of the total population admit to having played the online driving game at some time. And its peak usage reached 230,000 simultaneous drivers, a massive peak usage in only its second year, topped only slightly by the world record peak of *Lineage* which has reached 250,000 simultaneous users after seven years of existence.

What is it. *Kart Rider* was created by Nexon and is a very simple - and fun - race car game. The gaming controls are very simple. On the computer it uses only the CTRL, ALT and SHIFT keys, and the arrow keys. The game pits drivers against other live humans so it always has the realism of racing other people, rather than playing against computers such as the driving games typically on Playstation and other consoles.

It must be emphasized that *Kart Rider* is fun. It has similar element of Hanna & Barbera cartoons like *Tom & Jerry* and Looney Tunes cartoons *Wile E Coyote vs Roadrunner* and so forth. Cars can have missiles to shoot at the opponent. But you can have a balloon to climb above the missle when it is shot etc. Fun racing, not "realistic" racing.

The basic level of the game is free, online on computers. But for serious gamers there are ways to customize cars and purchase accessories etc to gain an advantage when racing against more experienced drivers. Like the phenomenon of dressing up one's avatar with brand label clothes and decorating one's miniroom with furniture, now heavy user *Kart Rider* gamers are buying better virtual cars to the game so they can enjoy a better racing experience. Nexon has more than 100 separate items of content all at very modest cost, such as a new paint job 40 cents, or a top-end car, 10 dollars.

Heavy users are called "Kart addicts". *Business Week* reported in June of 2005 that professional sponsored racers played 8 hours a day racing in *Kart Rider*. Major tournaments are sponsored by global brands such as Coca Cola with 50,000 dollars in prize money, broadcast on two cable TV channels.

A game for all ages, over half of *Kart Rider* gamers are over the age of 20 and a significant number have never played an online game prior to driving in Kart Rider. In 2006 *Kart Rider* was accepted as an offical e-sports entry by the Korea e-Sports Association.

Chapter XII
Consumer Robotics

One Into Every Korean Home

Image courtesy *IT Korea Journal*

"Progress is impossible without change, and those who cannot change their minds cannot change anything."
George Bernard Shaw

XII
Consumer Robotics
One into every Korean home

Japan is often considered the leading country in robotics. A good example is the Banryu, a household guard robot, which monitors the condition of the home, and in case of emergency such as a fire, the robot can use its camera and mobile telecoms connectivity to call the owner and send pictures of what is going on in the home. Similarly anyone entering the premises will be greeted by the robot - and an image again sent to the owner. A useful modern household gadget no doubt, but the Banryu costs about 18,000 dollars.

South Korea wants to catch up - and pass - Japan in the lead in consumer robotics. Thus it comes as no surprise that SK Telecom in South Korea has also introduced a home guarding robot, only this is just 50 cm tall (under two feet) and weighs only 12 kg (25 lbs), does all the Japanese model can do. But the Korean robot is sold for only 850 dollars. The competitiveness of the Koreans is illustrated by this example. This helps explain why the South Korean government fully expects Korean homes to have household robots within ten years.

A ROBOT MEANS I WORK

The origins of the word robot come from the Czech word "robota" for work. Most robots in the world are intended to replace menial work, but of course from the first Sony Aibo toy robot dog, robots can also be created to be entertainment, toys, "pets" and companions as well.

Starts with the plan

Like other key initiatives to achieve a digital Korea, the Robotics initiatives are a structured program with deadlines and strategic networks of skills in place, from government, R&D and commercial companies. Early commercialization, feedback from digital Korea citizens and competition will make Korea a powerful and competitive force in the advancement of robotics.

We should point out to readers of this book, that the vast majority of the millions of robots in use worldwide today bear no resemblance to the science fiction humanoid robots such as R2-D2 and C-3PO of the Star Wars movies. Today robotics already exist in areas from industrial manufacturing automation to military drones to space exploration. All of the recent probes to Mars can in fact be classified as applications of robotics as none of the probes to Mars has been manned. Another rapidly developing area is that of microbots and even nanobots used in medical experiments. This book is not intended to cover all applications of robotics. We will focus on looking at robots how they integrate with humans and digital society in South Korea today. Consumer robots. Within a very short time consumer robots will be playing an increasingly important role in normal life of everyday citizens, all around the world. South Korea expects to lead this transformation.

South Korean robotics

Currently the South Korean robot industry ranks 6th in the world in terms of the total population of robots in use. Most of these are for industrial use in the automobile, semiconductor, and ship building industries. Thirteen conglomerate companies and 110 small and medium businesses participate in the manufacturing of robots. The domestic market size is about 350 billion won (350 million USD), and the world market share is about 3%. However, South Korea aims to become the third largest robot manufacturing country by 2013, and hopes to capture 15% of the world market share. A particular focus for the South Korean robotics industry is robotic application services combined with IT, in order to stand out from other countries - and to capitalize on South Korea's lead in ubiquitous computing.

However, designing practical consumer robots is considerably more challenging that designing industrial robots. Oh Sang Rok, the Chief of the

Ministry of Information and Communications of South Korea explains in the IT Korea Journal:

> Broadly speaking, there are two types of robot: conventional industrial robots and service robots. Industrial robots perform jobs for humans in factories and perform relatively simple tasks by following pre-programmed instructions. Service robots, however, assist our daily lives. The problem is that pre-programming them is not easy because the environments of our daily lives are not fixed. In any case, our definition of an intelligent service robot is a machine that is able to interact with people and carry out the relevant services after identifying its surroundings.
>
> Oh Sang Rok, the Chief of the MIC Communications, quoted in IT Korea Journal July 2005

While capitalizing on South Korean strengths, the robotics initiatives are also facing considerable challenges to develop practical robots that can function in environments with humans and in areas that often change, such as a home where children's toys form obstacles and furniture can be moved, etc.

Hubo is perhaps the best-known humanoid robot in South Korea. Able to walk, slide and even turn while walking - this apparently is a difficult achievement in robotics, Hubo is also able to walk backwards. Earlier robots would walk in straight lines and then pivot and then again walk straight. While of great interest to robotics engineers, these kinds of steps in teaching robots to mimic humanoid behavior are well beyond the scope of this book. However, the uses of robots available today illustrate well how far the industry has already come.

B HOUSEHOLD ROBOTS

Most of the current crop of mass market robots are aimed at households, for basic cleaning and monitoring uses, to act as a companion to children, and also increasingly to help older members in society. Oh Sang Rok, of the MIC said, *"Social and economic needs for intelligent service robots to support people's daily lives are increasing with the advance of an aging society."* Let us examine a few of the current crop of robots.

Cleaning the home

Ottoro is a household robot optimized for cleaning duties. It might be considered a very advanced and evolved intelligent vacuum cleaner. Developed by Hanool Robotics, Ottoro is operated by a remote controller. Equipped with sensors and an artificial intelligence camera Ottoro learns its layout and then is able to recognize its position wherever in the house. Hanool Robotics says that Ottoro can clean a typical apartment in about 40 minutes and of course, incorporating intelligence, the robot actually develops an optimized cleaning path by using an 'Automap cleaning method.'

The Yujin Consortium has developed a household robot they call Jupiter. The robot has a high power Pentium processor and a multimedia interface. Rather than remote control, Jupiter responds to voice commands, so when called it senses the instruction from its built-in microphone and Jupiter approaches the owner. Following voice commands it can perform typical household duties such as monitoring the home, checking that the gas is switched off, or singing a song for the children. Jupiter connects via a network and can provide information, images, videos etc to its owner. Jupiter operates automatically and knows how to recharge itself.

Teaching babysitter

The Hanool Consortium has focused more on the school age children and built educational support for its networked household robot, Netoro. Netoro helps with homework and research for the children in the family, as well as providing multimedia entertainment and information. The human model has been the live-in nanny, both playing with the children and teaching them. Netoro can be connected with the teachers of the school via the internet for reinforcing lessons from class and homework. It also functions fully as a cleaning robot and home security robot, of course. Netoro also helps with morning chores, such as gathering weather and driving information that it provides to the adults in the family at the appropriate moments in the morning.

Networked humanoids

Most robots act independently, with a computer and a self-contained operating world. Mahru, developed by KAIST, is a networked humanoid robot. Mahru sends image, voice and sensor data to an external computer system through a wireless network, which allows both the networked central computer, and other connected Mahru robots to collaborate and to learn or adapt to situations. Mahru is rather advanced also in stand-alone mode, as it

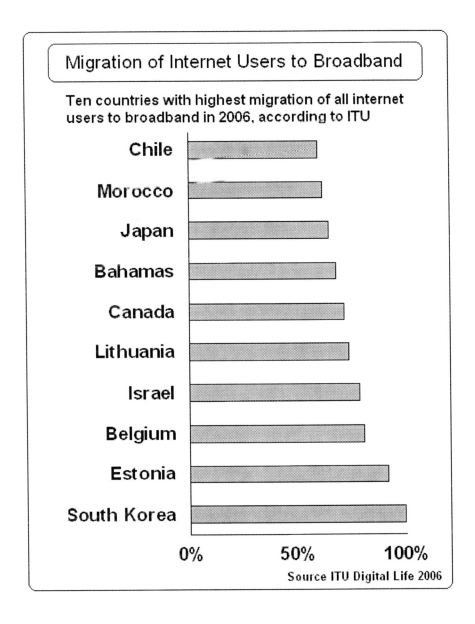

recognizes 100 voice-spoken words and sentences, senses obstacles with its artificial eyesight, and has the ability to chase objects. So for example if a Mahru gives instructions at a shopping mall to a person who is lost, Mahru can also then follow the person to make sure the human understood the instructions and went in the right direction.

Transformers come to life

Yujin Robotics has introduced a 2-leg walking transforming robot 'Transbot,' which transforms much like some children's toys, Transbot can participate in running games and combat games which makes it suitable as a companion for children who like to play fighting games and practice martial arts etc.

By 2005 there was a separate broadband connection for 18% of the American population
Source ITU 2006

C ASSISTANCE ROBOTS

Another common need for robots is in the various customer service areas where simple voice operation such as automation in calling center types of applications, is not enough. Robots can assist in anything from postal clerical duties to guiding shoppers to the right store.

May I help you to find your way?

Humanoid robots could be used in almost any instances of everyday human problems and issues. One is guidance, for example at shopping malls etc. Robotech has developed its Robo-X1 as the guide for humans. It uses Cartesian coordinates, voice and image sensors and access to databases. If you'd like to find the nearest store that has Levi's Jeans of a certain model and size, the robot could identify which store has that and then lead the way to guide you to the location. The same robot is used as the guide in restaurants to take you to your table, and in office buildings to lead you to the office or cubicle of the person you are meeting etc.

PGR, the "female" assistant robot

A novel approach to the humanoid assistance robot was taken by the Robotech Consortium, which aimed to develop a "female type" public assistant robot, which they call the PGR - Post Guide Robot.

This assistant robot will provide various services for public agencies such as the post office and various government agencies and departments. PGR does not have any security functions; it is purely a service device. In a postal assistant use, PGR can print out addresses and check on the status of a given postal shipment, but also provide a wide range of information such as biorhythms and even offer fortune telling.

> **By 2005 there was a separate broadband connection for 51% of the South Korean population**
> Source ITU 2006

D APPROACHING HUMANS

Science fiction has long promised a future of humanoid robots indistinguishable from human beings. Isaac Asimov's robot series of books set the benchmark for what to expect and most science fiction movies offer visions to inspire robotics engineers. The reality of how far humanoid robots have come may be quite surprising. Today many of the more difficult features of human replication have been solved.

Ubiquitous Robotic Companion URC

So do you want a personal loyal intelligent servant? The URC or Ubiquitous Robotic Companion is the vision of near term reality, where robotic companions will assist in information retrieval and information processing. URC offers entertainment functions and it handles many routine tasks. The

URC is enabled by exploiting the highly capable IT infrastructure of wireless and cellular networks of high speeds.

Using a cellphone from a remote location, the robot owner will be able to program URC for cleaning, monitoring for intruders, and gas valve checks. The robot offers various internet services, including entertainment functions such as reciting poetry and reading fairy tales for the children out of a large library online. For adults it can perform routine information retrieval tasks such as checking the incoming e-mail and reading it aloud. In addition, as it can reach the cellphones of its owners, URC can call up the owner and if there is a problem, send pictures and video and sound of what is happening, and await instructions from its owner. URC is still in development but is expected to be commercially available soon.

EveR-1 with humanoid skin

The EveR-1 robot in South Korea has managed to bridge the mechanical divide between metallic silver color tin robots to truly human-like appearances and the feel of real skin. EveR-1 has silicon skin and fifteen motors to mimic human movement. EveR-1 is formed to look like a Korean female in her 20s. The robot has sensors related to her eyes so that the robot can hold eye contact. The speech synthesis and logic allows EveR-1 to carry on a conversation and the robot's face can express emotions such as joy, sorrow and anger. EveR-1 is not fully mobile, she is intended to be sitting only, and has only an upper body. However, from a distance, she can easily be mistaken for a human female, sitting, discussing matters with people.

Bringing humanity to technology

As robotics make inroads into our lives, we will see a lot of conflict and clash, where humans meet technology. We do not mean a science fiction-esque robotic rebellion like in Isaac Asimov's I Robot series of books (and the movie starring Will Smith). But rather the daily interaction of technology. A good example is discussed by Mark Curtis in his book *Distraction: Being Human in the Digital Age*. Mr. Curtis discusses how we as humans insert humanity, our behavior and emotions, even in something as mechanical as typing:

> *Many people will be familiar with the problem of e-mail causing unintended offence. The brevity of the medium and the ease with which it is used sometimes cause real problems because recipients fail to distinguish the intended meaning clearly. Sarcasm or heavy*

handedness is often diagnosed when the writer simply was in a rush. Early users of e-mail understood this: and the result were emoticons such as :-) . I have not tracked the early history of these, but guess that they evolved from someone experimenting with keyboard combinations in order to express happiness, gloom, irony (my favorite – the raised eyebrow). It is interesting to note that hundreds of years of letter writing in the west had not developed emoticons. They came with digital.
　　　　Mark Curtis *Distraction* Futuretext 2005

　　　Similar examples will be emerging as we start to interact with robotics in our homes. Perhaps the industrial manufacturing robots, military robotic drones and space exploration robots can be seen so much as instruments, that we don't need to project our humanity upon them. But a home robot? Certainly we will start to personalize these, and invent "human-like" ways to cross the communication barriers between them and us. The current generation of consumer robots like EveR-1 will start to bridge that gap between humans and robots.

Toy Robots

Robotis demonstrated a robot kit named 'Bioloid,' through which you can learn about robots amusingly but without any special knowledge. The toy robot can be built as an excavator, or a dinosaur or a puppy. The kit also includes connectivity to a PC. The resulting robot can be controlled from the PC for simple things like basic movement.

> **Pearl - Robot and remote eyes.** Your robot calls you, plays video to your 3G cellphone. Most South Korean household and security robots include 3G cellular connectivity. If they discover something strange, they will call you on your 3G phone and show via live video camera what is going on.

　　　With 200 manufacturers we could not hope to catalog all South Korean robots but a few others warrant mentioning. One is the emotionally advanced robot from Iotek, which is intelligent in interpreting e-mails and able to engage in foreign language conversation to help teach languages. ETRI has developed its Wever series of robots that have very sophisticated pattern recognition scanners to and can detect their owners by facial features from a distance up to five meters (15 feet).

E A CULTURE OF ROBOTICS

South Korea approaches the goal or leadership in robotics very meticulously. In addition to the government initiatives, and already 200 companies working in this space, there are journals, conferences, trade shows and even competitions.

Intelligent Robot Exhibition

South Korea hosts an annual 'Intelligent Robot Exhibition' which in 2005 featured 30 computer manufacturing companies demonstrating their various robots. The Exhibition also features such supporting events as the Intelligent Robot Contest and the Robot Olympiad. The event gets a lot of publicity and as the South Korean robotics industry focuses strongly on consumer robotics, there often are fascinating robots with human-like appearance, behavior and features.

Robot Olympiad

To help foster development in robotic skills, the Robot Olympiad is held and it already attracts over 200 teams from South Korea and beyond in areas competed for the championship of robot labyrinth games, survival games, stair climbing, avoiding obstacles, emergency rescue work, robot football, and creative robots.
    The robot fight contest is a regular crowd favorite and a lot of prestige is involved in which robot team wins this tournament. Typically from one to two dozen teams enter humanoid style fighting robots. The contest runs a series of preliminary heats and then the finalists run head-to-head until a winner is determined.

Future

The competition for the development of personal robots is expected to accelerate. In the background of the keen competition between companies is the expectation that the scale of intelligent robot market will rapidly grow to

10 trillion won (10 billion dollars) by 2010 and 100 trillion won (100 billion dollars) by 2020.

You will be assimilated

This chapter has taken a quick survey of the state of the dawn of consumer robotics in the most connected information society on the planet. When computing and connectivity are ubiquitous, then consumer robotics are a natural evolution in how technology spreads into society. In addition, those who oversee the digital landscape of South Korea are in a particularly advantageous position to evaluate their impact. We turn to Oh Sang Rok, the Chief of the Ministry of Information and Communications in South Korea to put robotics into context. Oh Sang Rok says, "I believe the most innovative products that changed the 20th century are the PC and the Internet. What will we see in this century? I like to believe it will be robots."

Case Study 11
uPostMate

South Korea is one of the leading countries for robotics. Rather than the industrial, military, scientific etc robots, South Korea is focusing on consumer oriented robots to help in the home and in the daily lives of its citizens

Several advanced robots have been introduced for residential use and in various customer service applications such as guiding use etc. One of the innovative robots is the uPostMate which like its name suggests, is optimized for post office duties.

uPostMate was developed by a consortium of Samsung, Dasa Tech, Mayhill and Samkyung Hitech. uPostMate is designed to appear as a male postal worker.

Size and weight like a human

The robot, at 150 cm tall (5 feet) is similar to a mid height Korean male. The robot weighs 90 kg (190 lbs) also similar to a postal worker, meaning it does not need reinforced floors to work on.

uPostMate is programmed for various tasks needed in a post office from general office duties related in sorting post and moving packages and providing postal assistance. It is also designed for security work and business guidance functions and very post office specific duties such as zip code searches and printing out addresses.

uPostMate operates by voice commands and it also has a touch screen for input. The robot is networked and can do various information retrieval and processing functions from seeking information to providing entertainment content.

> Recognizing that the robot would be in the post office at night when the humans have gone home, the security functions are particularly emphasized. uPostMate can detect unauthorized entry to the office, generate an alarm and call its masters. The robot can naturally be fully integrated with security service providers.

Chapter XIII
Digital Convergence

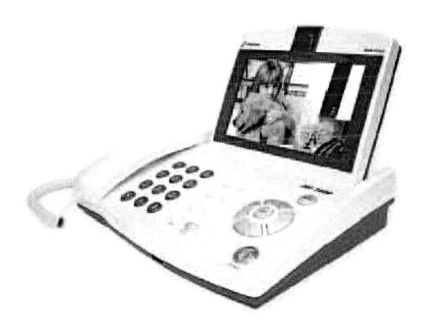

Internet, Telecoms and Media

Image courtesy *IT Korea Journal*

"Mobile phone companies like us will increasingly enter the media space as we see the need to add value to our customers."
Rick Kim, Head of Global Business at SK Communications

XIII
Digital Convergence
Internet, telecoms and media

Digital South Korea, the country with the highest penetration of broadband internet, highest usage of online videogaming, highest penetration of cameraphones, highest penetration of 3G advanced cellphones, the highest adoption of digital TV broadcasts to portable devices, and where already over half of all cellphone users make payments with their cellphone, replacing credit cards. For all these individual areas, there also is a convergence happening across these technologies and the industries they represent. For convergence in digital technologies, South Korea is years ahead of any other country.

A   Y OF CONVERGENCE

Many have discussed convergence of fixed and mobile telecoms, convergence of telecoms and the internet, and even the convergence of media and the web. Tomi Ahonen in his keynote to South Korea's largest digital convergence conference, iMobicon, in 2005, introduced his theory on the Y of Convergence that combines the elements, telecoms, internet and media. We discuss that theory here for first time in a book.

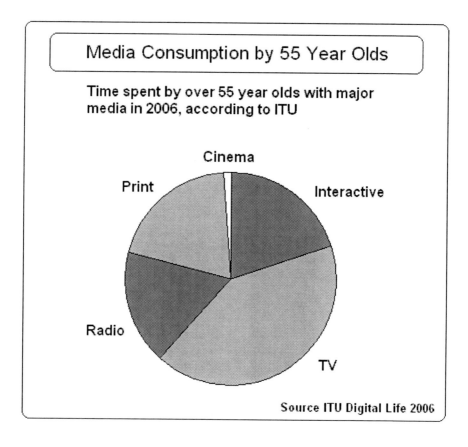

Three major elements converge

Like the letter "Y" there are three elements of technology - and their related industries - that form the major trends of digital convergence today. The three axes with no order of preference are datacoms, telecoms and broadcast. Each has several subgroups, but also each of the three axes has one dominant technology. With datacoms it is the internet - with about 1.1 billion users at the end of 2006; with telecoms it is cellphones with about 2.7 billion users; and with broadcast it is television with about 1.4 billion TV sets in use.

To explain the "Y of Convergence" we will be using those three dominant technologies as the proxies for the convergence. So in a significant simplification, but one that still holds the major elements of the theory, one

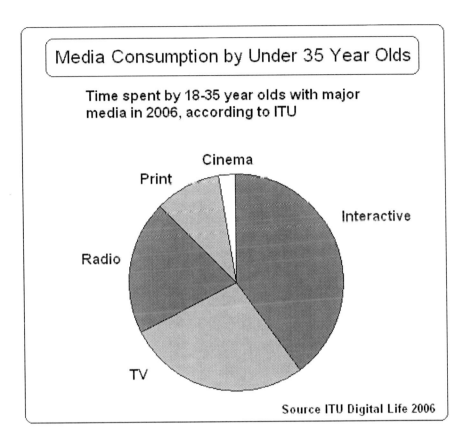

can say that digital convergence is the simultaneous merging of the internet, cellphones and TV.

Datacoms / the internet

The datacoms axis of the Y of Convergence is that leg which was always digital from the start. It was built without a commerce element but recently many technical solutions such as Paypal have been introduced to enable payments on datacoms networks. As an interactive media, datacoms could always resort to payment by credit card, and request the buyer enter card details on a form or in an e-mail message.

While some wireless networking technologies (so-called 802.11x standards) were being developed in the 1990s, it was not until the advent of

WiFi that datacoms and the internet started to become seriously wireless. There are about 1.1 billion internet users, of which about 750 million access on a paid PC based subscription, another 250 million or so access on a paid cellphone based subscription; and the rest are accessing via various campus networks like those at universities, internet cafes, etc.

Other datacoms networks tend to be also "IP Based" (using the Internet Protocol) but "closed" networks or ones with restricted access. There are still many proprietary and older networking technologies that also are used to connect datacoms devices. The internet is going through evolution to ever-faster forms of broadband speeds, as well as migration to wireless, and a technology upgrade to a new standard, IPv6 and a more collaborative internet called Web 2.0. Worldwide the leading countries where internet/datacoms innovation is taking place are South Korea, Hong Kong, Denmark and Sweden. The USA and Canada are both well ahead of the mainstream but no longer the leading industrialized countries on this axis.

3% of British cellphone owners have downloaded video to the phone

Source Telephia January 2007

South Korea is the obvious world leader in the internet, now with per capita penetration of broadband having passed 100%, and with all of internet users having migrated to broadband. Reflecting that leadership, South Korean broadband speeds are also the world's fastest, while the costs of broadband are the world's lowest.

Telecoms / cellphones

The telecoms axis has been going digital for the past two decades and currently almost all of the cellphone technologies are already digital. Cellphones are inherently wireless. Content appeared onto cellphones in the late 1990s and early in this decade, and commerce appeared roughly at the same time. The built-in payment abilities of cellphones were soon discovered on internet services and on TV interactivity, with many payment, voting etc solutions now using the payments of cellphones to "payment-enable" an internet application or TV programming, typically pioneered in both in gaming.

Finland was the first industrialized country of full fixed landline phone penetration, where cellphone penetrations grew past fixed landline phones in 1998. Since then practically all industrialized countries have seen the same phenomenon, in the USA this finally happened in 2005, and Canada is the only industrialized country left where this is yet to happen. By end of 2006 there were twice as many cellphone subscriptions as fixed landline phone connections.

The wireless/cellular industry is going through a migration to "next generation" networks often called "3G networks". The leading countries where cellphone innovation is taking place today are Hong Kong, Italy, Sweden and South Korea. Obviously the USA and Canada are the laggards by this axis of convergence.

While South Korea does not lead cellular telecoms in total

12% of South Korean cellphone owners have downloaded video to the phone
Source NIDA September 2005

---

subscription penetration, in terms of global leadership on the cutting edge of the technology, South Korea is again far ahead of all other countries. South Korea was the first country to reach 50% migration of cellphone customers from 2G to 3G. Currently all cellphones sold are cameraphones and over half of internet access originates from cellphones.

Broadcast / TV

While radio has more users, of the world's media consumption, television gets much more of our attention, and also as an industry TV dwarfs that of radio. Broadcast industries have transmitted content from the start, but only in the late 1990s have started to move to digital transmissions including IPTV and now are just launching digital wireless technologies. The digital technologies allow theoretically interactivity, although in most cases actual interactivity is rather done via the cellphone as in the American Idol voting etc. One should remember that over-the-air broadcast is wireless as such as is satellite broadcasting, but there are cable TV systems that are currently still mostly wireline based.

Broadcast did not have direct payment systems nor interactivity, but digital TV has introduced some of those abilities, depending on the technology and TV (and radio) is now strongly expanding programming

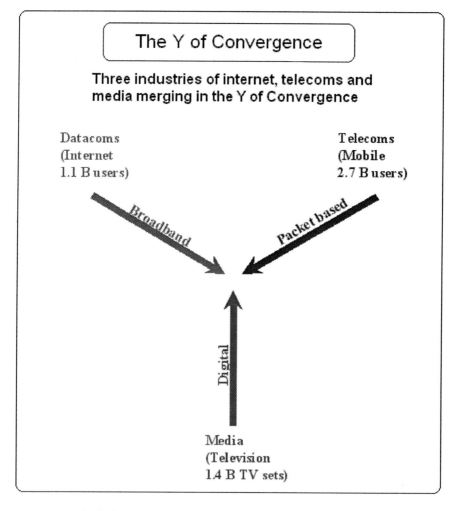

concepts to include viewer/listener interactivity and to a lesser extent also commerce. The TV industry in the USA and UK has recently seen the majority of industry revenues shift from advertising to subscription based. In Finland the TV industry has gone even further and today generates more revenues from cellphone interactivity than advertising and subscription revenues combined.

The TV and radio broadcast industries are migrating to digital technologies and are now introducing portable wireless technologies on standards like DVB-H, DMB and MediaFlow. The leading countries in TV innovation worldwide are the USA, Japan, Canada and the UK.

Again while USA and Japan lead in the total cable TV/satellite and digital TV industry, of its portable digital element (the cutting edge of TV broadcasting) South Korea is miles ahead of everybody else. In the new handheld digital TV technologies South Korea was first to launch both a Satellite and Terrestrial variant of portable digital TV technologies (i.e. building in your digital TV set-top box right into the cellphone handset). Since May 2005 and in just over 18 months they signed up more than three million customers to this type of broadcast mobile TV service, of which 2.2m on Terrestrial DMB and 0.85m Satellite DMB.

Center of convergence

At the center of the three axis of the Y of Convergence is the end-state, the focal point for all converging technologies. Approaching the point from the side of internet/datacoms we find evolutions of broadband wireless internet in the form of WiMax and WiBro as well as such voice-enabled datacoms technologies as VoIP (Voice over Internet Protocol).

The same kinds of needs are aimed to be met with 3.5G technologies from the cellphone side, with HSDPA the most prominent technology. Moreover, the digital TV industry is aiming to meet those needs with DVB-H, DMB (and DAB in radio broadcasts).

As the center of the convergence is inherently packet based and on the IP (Internet Protocol) standards, as it is all wireless, it is all broadband, and it all allows content delivery; the center is also a new battleground for the dominant players in each of the three converging elements.

In 2005 we saw many players staking claims in the other converging areas such as internet giant Google going for wireless; Disney launching cellphone operations as "MVNOs" (Mobile Virtual Network Operator) while the Japanese wireless carrier giant NTT DoCoMo has become the world's largest internet service provider solely based on its cellphone subscribers on its i-Mode internet service.

Currently the leading countries where the digitally converged world is most evolved are South Korea, Japan, Sweden and the UK. And South

Korea is by far the most advanced leading with the commercial availability and end-user adoption of the related technologies.

B CONVERGED SOLUTIONS

Another key driver for making convergence happen is the capability of the service providers to actually put together the digital packages and extra services of most interest. Now with high network speeds in South Korea, Japan, and other leading countries across ADSL/VDSL, Cable Modem and FTTH we find Digital TV maturing. The reduced price across LCD TV has fuelled a dramatic increase in the adoption of these necessary building blocks to lay the foundation for the converged IP based set platforms. Set top boxes are in the home for Home Networking, triple play services and wealth of VOD and other Multimedia on demand opportunities have emerged as well.

This has further enabled a reduction in the prices of managing the multimedia content at the network end of the communication system, from services to broadcasting. These advances have in some cases eliminated and in other cases postponed the need to upgrade network infrastructures.

All of the above highlight the success of SK Telecom (SKT) in particular, already famous as one of the most respected Wireless carriers in the world. It must also be emphasized that SKT is one of the most accomplished convergence players in the truest sense, combining its core voice and data business (Nate services for 2.5G and June for 3G Networks) with Satellite Mobile TV, Internet portals (Cyworld), Entertainment (Film and TV production and distribution and music labeling).

SK Telecom's offer is based on vertical collaboration at all levels as opposed to the typical service offers in Europe which tend to make the mistakes of a technology approach missing the life cycle view necessary for fast growth customer orientated services. Looking in more detail at the convergence of SKT wireless portal the following brands are key : Melon for Music, Cizle for cinema/VOD, Cyworld for interactive blogging ,M Bank and Moneta for Finance, Tu Media for Mobile TV/Radio (12 Video/26 Audio channels), and Nate drive for Telematics.

C BROADBAND CONVERGED NETWORK, BCN

Moving from Convergence to South Korea, let us put the theory into practice. BCN or the Broadband Converged Network is South Korea's initiative to deliver next generation converged network technology. In any network that is honestly converged there should be no difference between methods of access and the service that is delivered. BCN builds on broadband, WiBro, 3G cellular and DMB technologies to deliver converged solutions. Of course as these separate network technologies merge, there will be gaps and imperfections to the converged outcome.

> **In the summer of 2006 South Korea was one of only two countries where 50 Mbit/s broadband was commercially available**
> Source: Analysys October 2006

With BCN for the urban areas like Seoul, all four networks would be available to ensure coverage for high-density and high-capacity needs. In the most distant locations, such as in the sparsely populated forests and outlying islands only one or two of the four would be enough to give sufficient service reach and capacity. South Korea aims to become a leader in deploying the information society and is eager to provide know-how and lessons from its path to this future, and assist other nations in learning from the Digital Korea experience.

Fits the Y

As we showed in the Y of Convergence, each of the areas is merging and in the center of the Y all converged technologies are wireless, high speed, digital and based on the internet protocol. The broadband connectivity of the BCN are increasingly mobile or wireless and their reach expands ever further beyond the cities. Distinguishing boundaries between each technology are blurring and even fading. Already almost half of South Korea's total broadband connectivity is mobile or wireless on cellular networks such as CDMA 1x EVDO and WCDMA, and on wireless data networks such as WiFi and WiBro. South Korea is the first country where all of these wireless networks are in commercial production for the same region. Through convergence South Koreans have access to fast information and will have choices and options among the various network and service providers to pick the performance and price points they want for the usage they desire.

The vision for the converged network was previously known as the Next Generation Network (NGN) and this term still is used widely internationally for example in technology standardization work with the ITU-T. In South Korea however the new definition is BCN. The National Computerization Agency (NCA) of South Korea has defined the Broadband Converged Network, BCN as:

> *A Next Generation integrated network that is accessible anywhere without any connectivity problems while offering top-class security for quality streaming of broadband multimedia services in an info-communications environment that embodies the convergence of fixed line and wireless networks.*
> Source: NCA

The definition establishes the reach of the BCN. It is not only a next generation telecoms and datacoms network model, but also it embraces the broadcast networks up to terrestrial and satellite digital TV platforms. The BCN is seen as one vast, fast internet protocol based family of networks where users can easily access all information and entertainment.

Not easy

Convergence has been a recurring theme in the telecoms and IT industries for over ten years. Most of the solutions around convergence have been partial, at least when viewed from the comprehensive viewpoint of the BCN. The South Korean vision for BCN, however, is the first national vision for a future fully converged network that is also being developed by all of the necessary elements in the country. Combining so many distinct technologies with their particular needs does introduce a multitude of challenges from the technological to the commercial and even political. Who issues the licenses, are broadcast rules now applicable to internet and telecoms companies; how about banking industry safeguards, are they needed for e-commerce and m-commerce. What of rights around freedom of press, do they apply to bloggers and the online world, etc?

There are particular strengths and weaknesses for the various elements coming together in the BCN. For example video content such as movies and TV shows work best over broadband and digital broadcast technologies. Deep underground reach for data communications tends to work best with cellular technologies, such as reaching underground parking lots and subway trains.

Some services that have historically been exclusive to a given network, such as SMS text messaging, should move to more universal access. New service platforms or upgrades to existing ones may be needed as well as upgrades to billing systems or changes to how customers are charged. These types of changes usually introduce adjustments to the business models and such areas as termination rates between the various network operators (carriers) and service providers.

Technology vs. business models

The technology tends to be ready for the migration to BCN in South Korea. The current obstacles and delays tend to center more around the business models for the vision of the BCN. Some of the existing industry components that will be merging have established models of how their industries earn their money and these can come under severe threat from the vision of a grand converged BCN. For example internet content often does not rely on charging end-users, but rather collecting revenues from advertising. As similar content becomes available in "broadband" forms on handheld portable devices such as cellphones, it is likely that most advertisers will follow where the customer "eyeballs" are migrating. In the cellphone based mobile internet, most content is charged. Hence advertising brings an added "bonus" revenue stream for wireless carriers. Meanwhile the internet players feel a sudden decline in interest to advertise on their services.

Nespot Swing

This conflict between technology and business model was seen for example in Korea Telecom's demonstrations of its Nespot Swing service where users are able to seamlessly move between WiFi and other network technologies. It has been proven to work technically, but the business model implications have not yet been solved when our book went to print. The issues to be solved involved interconnection payment expectations between the different network operators and the existing structures and practices which would need to change, often drastically. It is likely to be a lengthy and complex process of bilateral negotiations between the various network operators in telecoms, internet and media.

Typical of the complexity of the challenges is the very nature of the issue. If considering how to compensate one network provider for access to its network by customers of another network, consider the pricing. It is not simply what price level one operator has versus another. Across the technologies, there are distinct differences even on the principle of how the service is charged. In broadband internet networks the pricing system tends to be a flat rate model "all you can eat". In cellular networks the charging model tends to be based per use, i.e. per-minute or per-message charging. How to fairly compensate the participants, and also how to reasonably charge end-users so as to not destroy viable businesses and promote arbitrage between various service variants is a complex issue indeed. These are being solved today in South Korea as the elements of the BCN are coming online.

D WIBRO

So you have WiFi (802.11x and/or W-LAN) and have heard of WiMax (802.16) just around the corner. What is this WiBro? It is what the South Koreans call the "Portable Internet". Let us put it into context by comparing it briefly to WiFi and WiMax.

Wireless and mobility

The WiBro Portable Internet has several advantages over WiFi for delivering broadband data wirelessly. WiFi currently has distance limitations to roughly 100 meters (330 feet) and does not support access by vehicles moving at speed. WiBro can be accessed up to a radius of 1-3 km (0.6 to 2 miles) around its base station and very significantly considering mobility, WiBro can be accessible at speeds of about 60-100 km/h (40-65 mph). WiMax will have access distances over a greater range, but WiMax is at least initially not specified to provide access from vehicles moving at speed.

     The carriers on cellular networks (3G providers) are particularly interested in WiBro Portable Internet technology because they already own and maintain the chains of cell sites, base stations and antennae which tend to fit density of a WiBro network. As WiBro also allows portable internet access from moving cars, WiBro high speed wireless data access provides a compelling complementary access method to 3G/3.5G cellular networks. The cellular networks are likely to become congested when data applications around video blogging, picture sharing, music downloading and multiplayer gaming expand the traffic loads from current loads that mostly are generated by voice and SMS text messagingp

Landline carriers

Similarly fixed landline telecoms carriers who do not own cellular licenses have monitored the shifts of traffic from the fixed network to cellular networks, and from fixed broadband internet access to mobile internet on 3G, and want to join in the wireless and mobility trends. WiBro offers a practical evolution path to deliver internet content and services wirelessly and with viable coverage for large areas that are not prohibitively expensive to deploy as would say a citywide WiFi network.

There is a technology gap between the 3G networks of cellular carriers i.e. CDMA 1x EVDO and WCDMA (known also as "UMTS") on the one hand, and WiFi and WiMax on the other hand. Cellular offers seamless data access over very wide areas and into fast moving vehicles. However, 3G is not fast enough to efficiently deliver broadband speeds to large populations. It is a voice communications-oriented technology and its data abilities are still compromised. WiFi and WiMax are high-speed data networks but they do not support moving vehicles, so they are really limited to hotspot use, for people who sit in a cafe and work for a half an hour or more. The need increasingly is the best of both.

Users appreciate mobility

A heavy data user wants high speeds like WiFi and WiMax but also does not want to limit data access only to times when being seated in a hotspot area. The users wants to continue connectedness - say listening to a news broadcast or streaming music station - and move to another location, take a taxi to a meeting and continue work on the collaboration site, etc. So the users want as much freedom of movement as the cellular network offers. In that way WiMax addresses this gap in high-speed data and mobility, making it a compelling offering for both sides of the industries that are now merging.

Solutions around this type of convergence are already being deployed commercially. KT introduced its One-Phone concept that allows automatic connectivity to the network which provides lowest cost. The One-Phone detects Bluetooth, WiFi, WiBro, WiMax, 2G and 3G networks and select whichever is cheapest for the given type of call. Handset makers Samsung and LG have both indicated they are ready to introduce handsets with these converged radio technologies the moment their customers, the wireless carriers (mobile operators) request them.

E CELLPHONES AND CONVERGENCE

The phone industry ships over 1 billion cellphones in 2007 and about 1500 separate phone models are in production at any one time. The industry innovates very rapidly with the big five - Nokia, Motorola, Samsung, LG and

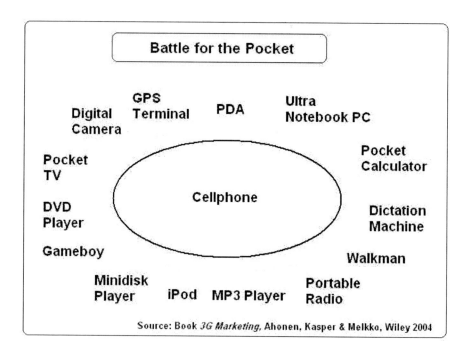

SonyEricsson - introducing numerous models annually, often several models per month. The replacement cycles for phones are 18 months worldwide and in the industrialized world over 20% of the population have more than one subscription, typically this means having two phones and an effective replacement rate of 9 months. Since Nokia first introduced the concept of phones being an industry like that of fashion, in 2000, today all major handset makers treat their handset stock as an inventory of rapidly diminishing value as fashions change.

Too fast

With this, it is almost impossible to write about the practical details about handset convergence in 2006 for a book that could be relevant still by end of 2007 or in 2008. The industry simply moves too fast. As we went to print Apple had joined the cellphone manufacturing business by announcing the iPhone and the very early buzz around the iPhone was that it radically shifts the whole industry. It is too early to tell but that does point out that radical innovation can happen in the industry and tastes can shift dramatically in only a matter of months.

Recent history has ample examples from design preferences (candy bar configurations vs. clamshell designs vs. slider phones) to consumer tastes in size. In Japan and South Korea a few years ago everybody was seeking the smallest phone sizes. Then with video and web content and now DMB broadcast TV, suddenly the screen size was the ruling size factor, and ever-larger phone started to gain popularity. The customer tastes in a fashion industry are extremely fickle. It is important also to bear in mind that heavy user, the youth and young adults, our Generation-C for the Community Generation, replace cellphones more frequently than they replace their Nike and Adidas sneakers.

The Y of Convergence and the BCN network philosophy do resonate strongly with handset convergence. The first voice call on a commercially launched cellular service was available in Japan by NTT (now NTT DoCoMo) in 1979. The first data service beyond voice calls is SMS text messaging which was launched in Finland by Radiolinja (now part of Elisa Group) in 1991. These are considered the classic cellular services, available universally on all networks, standards and phone models.

Internet and cellphones

The first full internet access from a cellphone was in Finland to the exclusive and almost prohibitively high-priced first Nokia Communicator in 1997. By 1999 NTT DoCoMo launched its radical i-Mode service and the full internet became accessible via mass-market cellphones in Japan. It was only in 2001 that the first ever TV video clips were available for smartphones in Finland from Finnish TV broadcaster MainosTV3, for the latest model of the Nokia Communicator at the time. At the same time in Japan J-Phone (since renamed Vodafone KK and then sold to Softbank) introduced its line of revolutionary cameraphones and its ShaMail picture sharing services. In 2003 in South Korea the first commercial MP3 downloads direct to musicphones were launched.

Still in 2003 there was no cellphone in the world that offered full internet access, TV video access, MP3 player and camera, integrated into one phone. There was convergence already in cellphones, but it was partial convergence. One phone might combine voice, SMS, web browsing and camera. Another might substitute the camera for an MP3 player. A third might substitute the MP3 player for video clip access.

Now all have it

By 2006 the top-end models from Nokia, Motorola, Samsung, LG and SonyEricsson all had cameras, web access, MP3 players, and video clip

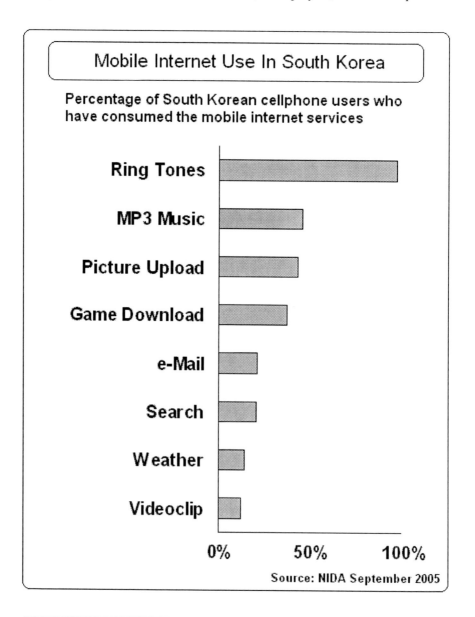

access. Some like the SonyEricsson Walkman series were optimized for music consumption, others like the Samsung SCH B600 is more of a top-end digital camera with its 10 megapixel resolution than a classic cameraphone. Still others like the Nokia N-93 were optimized for video consumption with DVD quality video recording and TV-out connectivity. Nevertheless, almost every model of the mid to top range for the top five suppliers included each of the technologies. Every one had a camera, MP3 player, web browser and video clip viewing ability. This massive amount of convergence happened in a few years while adding 3G and WiFi radio technologies and reducing overall handset size and increasing battery life.

The complexity inherent in cellphone design was summarized very well by Samsung's Senior Advisor Won Kim who is the Chair of the ACM Special Interest Group on Knowledge Discovery and Data Mining. Mr. Kim wrote in the Journal of Object Technology:

> *Japanese cell-phone makers estimate that the cost of developing a cellphone today is split 80 to 20 between software and hardware, while the split was 20 to 80 in 1998. Large US software makers now estimate that high-end embedded software will consist of up to 9 million lines of code in the near future. All digital devices are embedded systems. Some consist of hardware and firmware only, while others consist of hardware, firmware, and software burned into flash memory. High-end embedded systems, such as high-end cellphones and digital television, have large-scale complex software. As vendors of such devices are being pushed, by competitive force and their own drive to increase profits, to add more features and improve on current features, the devices are increasingly becoming software-oriented. The current competitive landscape requires vendors to churn out many models of the devices in ever-compressed production cycles. The current production cycle for many consumer-electronics devices is six months; and sometimes it is even 3- months.*
>
> - Won Kim, Samsung, writing in *Journal of Object Technology* Vol 4, Number 4, 2005

The needs in cellphone design and manufacturing therefore are increasingly computer-programming skills. While some parts of the handset are industrial manufacturing skills, such as the casing for the phone and its battery, most of the actual value of the phone is now in software. For this demanding area the world is facing a big shortage of skills. Some IT companies that specialize in cellphone design and user interfaces have emerged, most obviously Fjord in the UK which already does outsourced UI design work for leading handset makers and carriers (mobile operators).

PBX replacement

An interesting twist on the convergence story is corporate/enterprise business customers. They often want to provide mobility for their employees but do not want to pay cellphone charges for internal calls or calls originating from the office premises of the employers. Various technologies and commercial packages have attempted to address this need, from IP telephony (VOIP Voice Over Internet Protocol) like Vonage and Skype, to WiFi and converged phones, to the "Home Zone" pricing models offered by many pure mobile play wireless carriers against their fixed-mobile incumbent rivals, as was first launched in Denmark at the turn of the decade.

96.8% of South Korean handsets were mobile internet enabled by 2005
Source: Gartner Dataquest 2005

Now Samsung offers its solution to the needs of the enterprise, with what they call the W-PBX (Wireless Private Branch eXchange). With Samsung's solution there is a company-operated internal, private cellular network inside the office building(s) with tiny base stations with low power that operate only inside the building. Any calls that are between employees of that building will be handled by the private network and routed via the W-BPX to never connect with the cellular network provider networks. Only when the employees step into a taxi to go to a meeting or lunch outside of the building, do they connect to the cellular carrier's network and then legitimately generate billable cellphone calls. The particular benefit is that the employees can use real cellphones with all the stored numbers etc, and the employer need not lecture the staff about ignoring the fixed landline phones sitting idle on their desks.

Counters VoIP and WiFi

Many carriers with cellular network licenses are welcoming this technology, as it pulls the mat from under the VOIP and WiFi technology providers who are making inroads to the enterprise customer space. It also provides a compelling offering to fight against fixed landline operators (or the big incumbents in most traditional markets who have both a fixed and mobile network). The costs of managing the mini base stations inside the buildings are covered by the enterprise/business customer. These include installation, electricity, maintenance etc.

Now the employees are all equipped with cellphones, which will then generate a lot of traffic when the secretary for example has to rush home to care for the child who was sent home from school, etc. That secretary can still be connected via the cellular network and continue to work, but now the employer is happy to pay for the temporary increase in cellphone costs as without it the secretary would be out of reach for the rest of the day. Carriers are expecting this W-PBX solution would yield a significant about of additional incremental traffic not otherwise captured.

Convergence is so much more

We have touched upon the industries converging in the Y of Convergence, the networks converging in the BCN and discussed briefly how cellphone handset functionalities are converging. However, South Korea offers illustrations and commercial applications of so many other concepts around convergence. Money and the web are converging, with the Dotori, the acorn, money unit inside Cyworld. Humans and virtuality are converging in our avatars. TV content and user-generated content are converging as how Tu Media show user-generated videoclips on the air. Devices are converging from the household robot incorporating the cameraphone to send pictures and video to the homeowner, to the intelligent car.

Concluding Convergence

We cannot hope to cover all possible areas of convergence in just one chapter. Nevertheless, we hope we have given the reader a glimpse into this aspect of the future. Certainly every technology element discussed in any chapter in this book can also be seen to converge. That is the ultimate point to this chapter here at the end of the book.

But yes, gigabit speed internet access, already being deployed on the ETRI campus and to be launched wirelessly before this decade is over.

Certainly South Korea is the Mecca for digital convergence. If you are in any industry we have discussed in this book, you owe it to yourself to make your pilgrimage to Seoul this year and see the future of your business and its impacts to your personal career. South Korea is Digital Korea.

Case Study 12
Gigabit Broadband

Faster than a speeding broadband? South Korea has the world's highest penetration of broadband. It was the first country where all internet access had been migrated to broadband. It also has the world's lowest prices for broadband and the world's highest speeds for broadband. Yet this is not enough.

Now the South Korean ETRI Electronics and Telecoms Research Institute is working on what they call superfast internet, at speeds of 1 Gbps transmission speeds. Usually broadband speeds are understood to be in the one megabit per second speed.

Gigabit speed broadband is effectively one thousand times faster than basic broadband speeds currently rolled out in most countries.

How much faster than the current top speeds? Gigabit broadband would be about 10 to 20 times faster than the world's fastest commerically available speeds in South Korea and Japan today.

Superfast internet is based on FTTH (Fiber To The Home) technology and has already been deployed for 16 pilot customers at the ETRI campus in Taejon South Korea since 2006. ETRI expects commercial launch by late 2007 or early 2008.

How about mobile gigabit broadband?

Not to be outdone, Samsung is already showcasing its technology demonstration at Cheju Island in South Korea with wireless gigabit broadband. Samsung has already demonstrated 1 Gbps downlink speeds, although these are still only possible under optimal conditions as Samsung still develops this technology.

The solution was unveiled at the 4G Forum held in South Korea in 2006. Samsung's system supports 1 Gbps at standstill and 100 Mbps when connecting with users who are moving such as accessing wireless broadband from a moving car. Samsung expects the ultra-fast portable internet will be commercially available in by 2010 and hopes to show first prototype handsets for it by 2008. Yes, Digital Korea, it is only getting faster.

Chapter XIV
Conclusion

Can Korea Maintain its Lead?

Image courtesy *IT Korea Journal*

> *"In a connected age, sharing information is power."*
> **Tomi T Ahonen**

XIV
Conclusion
Can Korea Maintain its Lead?

Where is the future headed? We have seen South Korean statistics in this book and after a while it becomes mind-numbing to see the statistical lead. Let us put it in other ways. The rest of the world is still moving customers from narrowband (dial-up) internet access to broadband. That is typically at between 1 Mbit/s and 10 Mbit/s speeds. South Koreans currently are served at speeds from 50 Mbit/s to 100 Mbit/s, and by end of 2007 the rollout of Gigabit speeds (1,000 Mbit/s) will start to be offered.

In a very concrete way, South Korea is literally 100 times faster in the digital world than almost all of the rest of the industrialized world.

And access

But it is not only speed. Access. We have cable TV set-top boxes and satellite or digital TV in all industrialized countries. However, in South Korea almost 10% of the population already enjoy portable digital TV set-top boxes in their pockets in their cellphones and built into their cars and laptop computers. For TV executives this means totally different viewing habits. No longer is the classic prime time the only peak of the day. A new peak for viewing is lunchtime at work. In addition, a parallel viewing pattern emerges when homes have multiple digital TV access points. Not to mention time shifting of IPTV and PVRs.

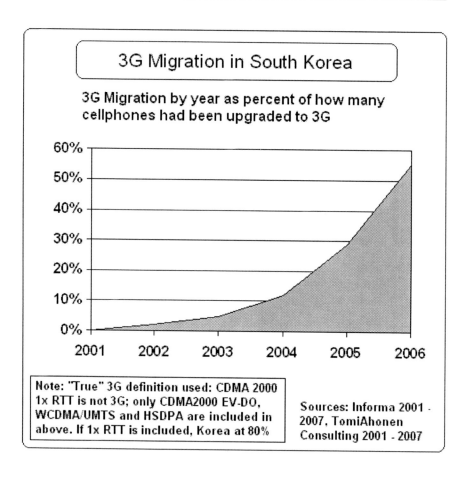

3G Cellphones and 2D barcodes

And then there are the 3G cameraphones. By end of 2006 nearly 70% of South Korean cellphone users had migrated to 3G. All of the phones are cameraphones, and almost all of them also have 2D barcode readers. Moreover, ease-of-use is vital in adoption of new technologies and in the change in behavior. We do see the 2D barcode revolution now spreading to Japan, as the major key to unlocking the internet power of portable devices that has for a decade been promised from PDAs, WAP phones and 3G phones. With over half of the population already using 2D barcode readers, South Korea shows that now finally we have cracked that difficult issue of convenience. Like KW Park explained:

> *"The 2D barcode opens the mobile phone browser to all web content. This will happen all around the world, it is that much of an improvement over the past. And from a technical point of view, the 2D barcode reader is just software. If you have a cameraphone, it can be done with minimal technical and cost penalties."*
> KW Park, COO Iconlab

Some of the latest innovations, like virtual worlds, digital TV broadcasts to cellphones, and indeed 2D barcodes, are likely to produce their own revolutions, first in Digital Korea and then around the world.

What of Japan?

One question we often get is, are Japan and South Korea similar. Certainly there are many similarities, but recently South Korean IT, telecoms and media companies have been remarkably successful in turning their local innovations into global success. We put the question to Lars Cosh-Ishii the co-founder of English Language observers of the Japanese telecoms market, Wireless Watch Japan who also closely monitor the South Korean IT industry. Mr. Cosh-Ishii explained the difference in international success in these words:

> *"One interesting comparison between handset makers in Korea and Japan is the demonstrated level of success they have achieved in sales over-seas. While the dozen different Japanese OEM's have had rather limited success, both Samsung and LG have clearly performed very well. There are certainly many factors to explain these results; however two immediate points stand out. First is the size of the respective domestic markets; having double the population, and relative affluence, Japan initially allowed companies here to maintain comfortable returns without 'having' to consider an aggressive global growth strategy. The second significant difference is so-called economies of scale; the Japanese tend to run wider model selections with shorter production lines and rely solely on the operators' sales channel for distribution while the Koreans adopted a simple low cost, large volume, approach to target still developing major international markets."*
> Lars Cosh-Ishii, Co-Founder Wireless Watch Japan

So in South Korea we see a society that is most connected, where digital is truly ubiquitous. We also see a country where the national ethos keeps urging all to Bballi bballi, to hurry-hurry. In addition, we see that the domestic South Korean market is not enough, and the companies soon set their eyes abroad, for expansion. Whether Ohmy News launching in Japan, Lineage spreading to China and Hong Kong or Cyworld appearing on American shores, and the LG Prada "iPhone clone" cellphone releasing in Europe before the iPhone is even launched, Digital Korea is also rapidly coming to a home near you.

It's been our pleasure

So we hope you have enjoyed our journey to Digital Korea. You may think that the issues we discuss in this book are only a fleeting technical lead, which cannot last. That other advanced high tech economies and countries, such as Japan, Singapore, Hong Kong, USA (West Coast in particular), Scandinavia, Italy, Israel, etc will soon catch up with South Korea. Here we want to quote one of our interviewees, who understandably asked to be anonymous for this comment. The Western IT executive met with South Korean technology partners and they discussed collaboration on a project. When the Western executive had provided his most aggressive project schedule to the South Koreans, the Koreans responded "You Western people! You don't move too fast." They were severely disappointed at what in South Korea is the expectation for moving rapidly.

It is that concept of Bballi bballi as we have discussed in this book. Hurry-hurry. It is what brought South Korea ahead of the rest of the digital world. We are convinced that they will Bballi bballi into the next decade, holding on to their lead. Therefore, our prediction - no other society will catch up to Digital Korea any time soon.

"Somewhere, something incredible is waiting to be known."
Carl Sagan

Abbreviations

2D	Two Dimensional
2G	Second Generation (current digital) mobile telephony, first launched 1991 in Finland by Radiolinja/Elisa. These include GSM, CDMA, TDMA etc
2.5G	"Two point Five G" enhancements to the second generation mobile telephony enhanced beyond 2G but that are not 3G. These include technical specifications GPRS, EDGE and CDMA2000 1x
3G	Third Generation (new generation) mobile telephony, first launched 2001 in Japan by NTT DoCoMo. Also known by the technical standard IMT 2000. Mainly "true" 3G consist of WCDMA (UMTS), CDMA2000 EV-DO, and TD-SCDMA
3GSM	Third Generation and GSM (Global System for Mobile)
3.5G	"Three point Five G" enhancements to the third generation of mobile telephony beyond 3G but that are not 4G. For example HSDPA
4G	Fourth Generation (next generation) mobile telephony, also known as "Systems Beyond IMT 2000" currently being standardized with spectrum to be allocated at World Radio Congress in 2007 and standard completed after that; with first commercial 4G standard-compliant networks expected to be launched 2012.
7 Mass Media	(also 7th Mass Media) seven mass media in order of their launch: Print, recordings, cinema, radio, TV, internet and mobile
802.11	IEEE standard for wireless connectivity, see W-LAN and WiFi
802.16	IEEE standard for broadband wireless, see WiMax
802.20	IEEE standard for broadband wireless

ADSL	Asyncrhonous Digital Subscriber Line (a type of broadband)
APEC	Asia-Pacific Economic Cooperation
ATM	Automated Teller Machine (cash machine)
BBC	British Broadcasting Company
BCN	Broadband Converged Network
Blog	weB log
CBS	Colombia Broadcasting Service
CD	Compact Disc
CDMA	Code Division Multiple Access (a 2G cellular telecoms standard)
CNN	Cable Network News
CRM	Customer Relationship Management
CTIA	Cellular Telecoms & Internet Association
DAB	Digital Audio Broadcast
DJ	Disc Jockey
DMB	Digital Media Broadcasting
DRM	Digital Rights Management
DVB-H	Digital Video Broadcasting - Handheld
DVD	Digital Versatile Disc
EA	Entertainment Arts
ESPN	Entertainment and Sports Programing Network
ETRI	Electronics and Telecommunications Research Institute
FT	Financial Times
FTTH	Fiber To The Home
HBO	Home Box Office
GDP	Gross Domestic Product
Gen-C	Generation-C (Community)
GM	General Motors
GPS	Global Positioning Satellite
GSM	Groupe Special Mobile, Global System for Mobile

	communications
HDTV	High Definition TeleVision
ICT	Information and Communication Technology
ID	IDentity
IM	Instant Messaging
IP	Internet Protocol, also Intellectual Property
IPTV	Internet Protocol TeleVision
IPv6	Internet Protocol version 6
ISDN	Integrated Services Digital Number
ISP	Internet Service Provider
IT	Information Technology
ITU	International Telecommunications Union
KBS	Korea Broadcasting System
KIICA	Korea ICT International Cooperation Agency
KIPA	Korea IT Promotion Agency
KISDI	Korea Information Strategy Development Institute
KMPS	Korea Mobile Payment System
KT	Korea Telecom
LAN	Local Area Network
LBS	Location-Based Service
LCD	Liquid Crystal Display
MDA	Mobile Data Association
MIC	Ministry of Information and Communication
MIDI	Musical Instrument Digital Interface
MIT	Massachussetts Institute of Technology
Moblogging	Mobile web logging
MMOG	Massively Multiplayer Online Game
MMORPG	Massively Multiplayer Online Role-Playing Game
MMS	Multimedia Messaging Service
MP3	MPEG-2 Layer 3 (Motion Picture Experts Group)
MPEG	Motion Picture Experts Group

MSN	MicroSoft Network
MTV	Music TV
MVNO	Mobile Virtual Network Operator
NFC	Near-Field Communications
NIDA	National Internet Development Agency
PBX	Public Branch eXchange
PC	Personal Computer
PDA	Personal Digital Assistant
PR	Public Relations
PVR	Personal Video Recorder
RF	Radio Frequency
RFID	Radio Frequency IDentification
RSS	Really Simple Syndication
SIM	Subscriber Identity Module
SMS	Short Message Service
TFT	Thin Film Transistor
TIM	Telecom Italia Mobile
UI	User Interface
UKP	United Kingdom Pounds
URC	Ubiquitous Robotic Companion
URL	Uniform Resource Locator
USB	Universal Serial Bus
USN	Ubiquitous Sensor Networks
VCD	Video Compact Disc
VCR	Video Cassetter Recorder
VDSL	Very high speed Digital Subscriber Line
VHS	Video Home System
VJ	Video Jockey

VOD	Video On Demand
VoIP	Voice over Internet Protocol
WAP	Wireless Application Protocol
WCDMA	Wideband Code Division Multiple Access (a 3G cellular technology)
WiBro	Wireless Broadband
WiFi	Wireless Fidelity (see also 802.11 or W-LAN)
WiMax	Worldwide Interoperability for Microwave Access (see also 802.16)
W-LAN	Wireless Local Area Network (see also WiFi and 802.11)
WWW	WorldWide Web
xDSL	(various versions of) Digital Subscriber Line

> *"Nobody is as clever as everybody."*
> Alan Moore, CEO of SMLXL

Bibliography

Ahonen Tomi. **m Profits: Making Money from 3G**, Wiley, 2002, 360 pp

Ahonen Tomi, Barrett Joe. **Services for UMTS: Creating Killer Applications in 3G**, Wiley, 2002, 373 pp

Ahonen Tomi, Kasper Timo, Melkko Sara. **3G Marketing: Communities and Strategic Partnerships**, Wiley, 2004, 333 pp

Ahonen Tomi, Moore Alan. **Communities Dominate Brands: Business and Marketing Challenges for the 21st Century**. Futuretext, 2005, 274 pp

Anderson Chris. **The Long Tail.** Random House, 2006, 256 pp

Baldry Maggie. **Digital Eves**. Futuretext, 2007, 177 pp

Bayler Michael, Stoughton David. **Promiscuous Customers**, Capstone, 2001, 256 pp

Beck John, Wade Mitchell. **Got Game**, Harvard Business School Press, 2004, 208 pp

Benkler Yochai. **Wealth of Networks**, Yale University Press, 2006. 528 pp

Calvo Agustin. **Open Your Eyes and Wake Up Your Business**, Agustin Calvo 2006, 30 pp

Curtis Mark. **Distraction**, Futuretext London 2005, 222 pp

Cohen Adam. **The Perfect Store. Inside eBay** Piatkus 2002, 332 pp

Donaton Scott. **Madison & Vine**. McGraw-Hill, 2004, 240 pp

Friedman Thomas. **The World is Flat,** Penguin, 2006, 624 pp

Florida, Richard. **The Rise of the Creative Class and How it's Transforming Work, Leisure, Community and Everyday Life**, Basic Books, 2002, 404 pp

Frengle Nick. **i-Mode, A Primer**, M&T Books, 2002, 485 pp

Gladwell Malcolm. **The Tipping Point: How little things can make a big difference**, Back Bay Books, 2002, 304 pp

Golding Paul. **Next Generation Wireless Applications**, Wiley, 2004, 588 pp

Grant John. **After Image. Mind Altering Marketing**, Profile, 2002, 270 pp

Hannula Ilkka, Linturi Risto. **100 Phenomena**, Yritysmikrot 1998, 212 pp

Jaokar Ajit, Fish Tony. **Mobile Web 2.0**, Futuretext, 2006, 176 pp

Jaokar Ajit, Fish Tony. **Open Gardens: Innovator's Guide to the Mobile Industry**, Futuretext, 2004, 176 pp

KISDI (Korea Information Strategy Development Institute). **IT Industry Outlook Korea 2005 Report**. KISDI, 2005.

Kopomaa, Timo. **City in your Pocket, the birth of the information society**, Helsinki: Gaudeamus, 2000, 143 pp

May Paul. **Mobile Commerce**, Cambridge University Press, 2001, 302 pp

McLelland Stephen. **Ultimate Telecom Futures**, Horizon House, 2002, 232 pp

Moore, Geoffrey. **Crossing the Chasm, revised edition**, Capstone Publishing, 2002, 256 pp

Pachter Marc, Landry Charles. **Culture at the Crossroads**, Comedia 2001

Radhakrishnan Rakesh. **Identity and Security**. Futuretext, 2007, 418 pp

Raymond Martin. **The Tomorrow People**, Financial Times Management, 2003, 279 pp

Rheingold Howard. **Smart Mobs: The next social revolution**, Basic, 2002, 288 pp

Rifkin Jeremy. **Age of Access**, Tarcher, 2001, 320 pp

Scoble Robert. **Naked Conversations: How blogs are changing the way businesses talk with customers.** Hungry Minds, 2006, 251 pp

Searls Doc, Weinberger David. **Cluetrain Manifesto: End of business as usual**, Perseus, 2001, 190 pp

Stiglitz Joseph. **Globalization and its Discontents,** W.W. Norton, 2003, 304 pp

Weiss Tom. **Mobile Strategies,** futuretext, 2006, 186 pp

Willmott Michael, William Nelson. **Complicated Lives Sophisticated Consumers; intricate lifestyles simple solutions**, Wiley, 2003, 260 pp

"Most of us don't know exactly what we want, but we're pretty sure we don't have it"
Alfred E Neuman (Mad Magazine)

Recommended Websites

Cyworld USA site
http://us.cyworld.com/

Dynamic IT Korea (KIICA English)
http://www.dynamicitkorea.org/aboutus/aboutus.jsp

Dynamic Korea
http://english.tour2korea.com

ETRI   Electronics and Telecommunications Research Institute
http://www.etri.re.kr/www_05/e_etri/

IPTV and Virtual Broadcast Studio
http://www.darim.com

KAIT   Korea Association of Information & Telecommunication
http://www.kait.or.kr/

Kart Rider
http://kart.nexon.net/

KESA   Korea Entertainment System Industry Association
http://www.game.or.kr/

KIBA Korea contents Industry & Business Association
http://www.kiba.or.kr/

KIBWA Korea IT Business Women's Association
http://www.kibwa.org/

KIICA Korea ICT International Cooperation Agency
http://www.kiica.or.kr/

Kinternet
http://www.kinternet.org/

KIPA Korean IT Promotion Agency
http://www.kipa.or.kr/

KISIA Korea Information Security Industry Association
http://www.kisia.or.kr/new/

KITA Korean International Trade Association
http://global.kita.net/

FKII The Federation of Korean Information Industries
http://www.fkii.or.kr/

KNRA Korea Network Research Association
http://www.knra.or.kr/

Korea ICT Europe
http://www.iparklondon.com

Korea IT SME & Venture Business Association
http://www.picca.or.kr/main.do

Korea Radio Promotion Association
http://www.rapa.or.kr/

Korea Times
http://www.koreatimes.co.kr

Korea Radio Promotion Association
http://www.rapa.or.kr/

Korea IT Times
http://www.ittimes.co.kr

KOVWA Korea Venture Business Women's Association
http://www.kipo.go.kr/kpo2/ek/index.jsp

KSIA Korea Software Industry Association
http://english.sw.or.kr/

KTOA Korea Telcommunications Operators Association
http://www.ktoa.or.kr/english

KTNET Korea Trade Network
http://homepage.ktnet.co.kr/ktnet

Lineage 2
http://www.lineage2.com/

MIC **Ministry of Information and Communication**
http://www.mic.go.kr/

NSO **Korea National Statistical Office**
http://www.nso.go.kr/eng2006/emain/index.html

Ohmy News
http://english.ohmynews.com/

SMBA Small and Medium Business Administration
http://www.smba.go.kr/main/english/index.jsp

Telecoms Korea
http://www.telecomskorea.com

TomiAhonen Consulting
http://www.tomiahonen.com

NIC (National Internet Development Agency of Korea)
http://www.knra.or.kr/

Ubiquitous dream hall
http://www.u-dream.or.kr/eng/main.asp

> *"To apply media laws on bloggers is like firing a howitzer at a swarm of mosquitoes. You will change the configuration for about half a minute, then it will be like you never fired it at all."*
> **Randolf Kluver, Singapore Internet Research Centre**

Recommended Blogsites
(Korea and Digital Convergence related blogsites in English)

Communities Dominate Blogsite
http://www.communitiesdominate.com

Daily Kimichi
http://thedailykimchi.blogspot.com

East Asia Strategist
http://eastrategist.com

KoreaCrunch
http://www.koreacrunch.com

Next Gen Web
http://blog.webservices.or.kr

Open Gardens
http://opengardensblog.futuretext.com

Plus 8 Star
http://www.plus8star.com

Smart Mobs
http://www.smartmobs.com

Web 2.0 Asia
http://www.web20asia.com

Xfiniti
http://weblog.xfiniti.com/ceo

Index

2D Barcodes 8, 26, 68, 110, 148, 176, 187, 250
3G 3, 10, 41, 48, 60, 78, 89, 121, 147, 150, 155, 160, 167, 173, 179, 191, 217, 232, 236, 250
3G Marketing 238
3GSM 99
3.5G 4, 169
7 Mass Media 80
7-Eleven 86
24 89
52 Street 166
100 Phenomena 59
160 characters 117, 146
802.11 (see also WiFi) 227
802.16 102

acorn (dotori) 35, 44, 243
acupuncture 107
Adamson, Walter 143
Adidas 239
ADSL 174, 232
Adult entertainment 157
advertising 81, 84, 111, 147
Africa 85
Ahonen, Tomi 17, 182, 225, 238, 249
Aibo 210
airline 50, 101
alarm 25, 60
alcohol and cellphones 24, 122
Almes, Guy 173
all-you-can-eat 236
alphabet (Korean) 32
Amazon 34, 41, 107
American Express 97
American Idol 229
Amp'd 86
Analysys 67, 233
APEC 76
Apple 9, 64, 79, 155, 159

arcade 28
Arirang 76
Asia 18, 32, 35, 42, 50, 85, 117
Asimov, Isaac 216
Aston Martin 39
ATM 100
auction sites 34, 103
Auction.co.kr 103
Audiosound 141
Australia 175
Austria 6, 121, 175
automobile telematics 137
Autonet 140, 144
avatar 34, 40, 101, 128
Axess Telecom 138

background music 44, 165
Bahamas 213
banking 97, 108, 122
Bank On 98
Banryu 209
bar 19, 41
barcodes see 2D barcodes
bathroom (use cellphone in) 22
battery 147
battle for the pocket 238
battle tank 51
bballi bballi (hurry hurry) 5, 11, 103, 174, 183, 252
BBC 79
BCN 232
BDDO 25
Beatbox 166
Bebo 4, 43, 48, 193
Becoming Buddies 49
Beckham, David 85
bed 23, 25
Belgium 19, 67, 174, 213
Berlin 89, 164
Betamax 82
Big Brother 90

biometric 63
bio shirt 66
Bioloid 217
billboard 102
B-kyung System 60
Blackberry 20
blog (blogging) 4, 11, 32, 46, 84
Bluetooth 62, 236
BMW 39
BOA 32, 167
bored 22
Borg (Star Trek) 17
Bowlingual 185
boyfriend 18, 52, 200
brand 23, 32, 35, 40
Brazil 62, 199
Britain/British see UK
broadband 3, 6, 10, 41, 44, 60, 67, 143, 173, 192, 202, 225, 245
broadcast 75, 81, 91, 230
Bronfman, Edgar 160
browsing 87
Brunei 175
bulletin board 69, 148
bus stop, intelligent 68
Business Week 99, 103, 106, 205
buyer beware 105

C-3PO 210
cable modem 232
cable TV 91, 202, 231
camera 5, 25, 29, 126, 238
cameraphone 3, 10, 26, 48, 110, 116, 142, 181, 238
Canada 6, 18, 44, 62, 67, 86, 133, 174, 180, 203, 213, 228
Canas 140
candybar 239
car 137
car and TV 126
car dealer 19
car telematics 137
Caraeff, Rio 160
Carphone Warehouse 19, 20, 159
Carpoint 141

cartoon 40
cash loan 102
casino chips 146
casual gaming 199
CAT 102
CBS 182
CDMA 98, 143, 180, 232, 249
celebrity 82, 191
cellphones 3, 18, 22, 27, 30, 34, 41, 46, 52, 59, 66, 78, 116, 127, 137, 143, 150, 181, 238, 240
cellphone
  as toy 24, 200
  banking 97
  camera see cameraphone
  communities 34, 240
  credit cards 98
  email 20, 240
  gaming 39, 47, 157, 191, 197, 238, 240
  internet 82, 137, 157, 177, 240
  learning 7, 51, 200
  mass media 80, 225
  music 44, 79, 157, 162, 238, 240
  payment see m-payment
  remote control 60, 216
  replacement cycle 24
  rules of conduct 184
  search 40, 240
  traffic cam 150
  translates 185
  TV, video 50, 75, 88, 126, 157, 228, 238, 240

Celrun TV 91
championship gaming 200
Chaplin, Charlie 81
charge card 98
chat 22, 40, 50, 84
cheating 11, 18, 195
Cheju 245
children 22, 46
Chile 202, 213
Chin, Daeje 76

Index

China 32, 42, 50, 54, 84, 137, 158, 199
Cho, Dae Hui "FoV" 200
Chun, Jung Hee "Sweet" 200
Chungbuk University 182
Churchill, Winston 3
cigarettes 23, 27
cinema (movies, Hollywood) 3, 9, 81, 194, 225
Cisco 179
Citizen 65
citizen journalism 84, 107, 115, 132
Cltus 143
City in Your Pocket, The 28
Cizle 232
clamshell 239
clothing, intelligent 63
CNN 88
Coca Cola 50, 205
Cohen, Ted 160
Color Ring 156
Commax 60
Communicator 239
communities (see also tribes, social networking) 3, 27, 31, 104, 182
Communities Dominate Brands 17, 182
computer (see also PC and laptop) 29, 31, 50, 59, 63, 118, 191
comScore Metrix 62, 82
concert 89
Confucius 108
Connected Age 5
console gaming 47, 191
consumer robotics see robotics
content (digital) 44, 50, 84, 91
convergence 4, 8, 225
copyright 83
Cosh-Ishii, Lars 251
CounterStrike 7, 41, 100, 195, 200
Cowon 166
Crazy Frog 156
credit card 31, 78, 97
crime 126
Croatia 144, 198

CTIA 22, 183
Cuba 198
culture 32, 178
Curtis, Mark 65, 216
cyber crime see crime
Cyworld 7, 10, 29, 30, 34, 39, 43, 46, 85, 100, 165, 182, 193, 232
Czech 209

DAB 231
Daewoo 41, 178
dance tutor 51, 164
Dasa Tech 220
data protection 122
dating 24, 41, 46, 51
Daum 167
Denmark 6, 62, 121, 228, 242
diabetes 107, 123
digital camera see cameraphone
digital fatigue 52
Digital Investor 143
Digital Life 213
digital music see music
digital rights 167
digital society 118
digital TV see TV
digital wallet 63, 97
Disney 86, 231
dispersed computing 64
display 187
Distraction 65, 216
DJ 82
DMB 60, 75, 86, 92, 126, 139, 147, 167, 179, 231
dog to human translation 185
Donkey Kong 28
Doom 28
Dosiak 166
DRM see digital rights
dotori 35, 44, 243
Dungeons and Dragons 54, 195
DVB-H 231
DVD 93, 164, 238

EA 191
ear ring MP3 player 64

eBay 34, 41, 52, 97, 196
EBU Technical Review 77
e-cash see digital wallet
Ecuador 198
education see learning
EF Sonata 138
e-government 115
Einstein, Albert 59
Electronic Arts 191
electronic government see e-government
electronics (home) 5
Elisa Group 155, 239
Elvis 81
email 21, 87, 140, 182
emergency 141
EMI 160
emoticons 217
emotional messenger 50
employees 118
English see UK
Enjoymap Moti 143
entertainment 191
ESPN 86
Estonia 174, 213
ethernet 174
etiquette 123
ETRI 218, 243
Europe 50, 106, 161
Everquest 41, 195
Europe 18, 82, 85, 89, 97, 117
EveR-1 216
Everquest 7, 54, 194
Ezon 63

family 108
fantasy leagues 51
farming for videogame gold 52, 119, 201
fashion 32, 40, 63, 238
fatigue 52
fax 21
Felica 101
female robot 215
Ferrari 39

fiber to the home 245
film (camera) 25
Final Fantasy 195
Financial Times 198
Fine Digital 141
FineDrive 141
Finland 6, 35, 51, 62, 79, 88, 92, 109, 155, 167, 179, 229
Fish, Tony 163, 193
fixed landline see wireline
Fjord 242
flat rate 236
Flickr 10, 34, 41, 85, 193
Flight Simulator 51
flirting 24
floor (intelligent) 123
Florin, Gerhard 191
florist 102, 108
football see soccer
footprint 105
Ford 50
"Fov" 200
France 63, 67, 199
friends 49
FTTH 245
Fujimoto, Dr 19
Funcake 166
furniture 42

gambling 157
games (also gaming) 24, 27, 32, 39, 47, 51, 83, 117, 147, 191, 197, 229
game show 82
Gameboy 238
gaming farmers 52
gaming hall 11
Gartner Dataquest 242
gasoline (petrol) 97
Gates, Bill 155
GDP 198
Gen-C see Generation C
General Motors 139
Generation C 7, 17, 28, 192, 239

Germany 50, 62, 64, 67, 87, 158, 175, 197
gift (digital) 45, 50
gigabit broadband 5, 245, 249
Gilbeot 141
girlfriend 18, 52, 200
GM 139
goggles 64
gold farmer 52, 201
Google 8, 23, 79, 83, 137, 231
Gotham Racing 200
GPS 137, 238
Grand Theft Auto 51, 191
Grandeur XG 138
group call 24
GSM 98
Gucci 40

Habbo Hotel 4, 34, 39, 85, 100
hacking 127
haggling 19, 103
handset subsidies see subsidies
Hanna & Barbera 204
Hannula and Linturi 59
Hallyu 32
Hana TV 91
Hanaro Telecom 91
Hanool Robotics 212
hard disk drive (TV recorder) 82
Hawke, Tony 191
HDTV 174
Head and Body 89
Health 123
Healthapia 123
Helio 49, 50, 86
Hello Kitty 35, 42
Helsinki 79
Herald Tribune 39, 49
Hollywood see cinema
home, intelligent 59
Home Zone 242
Homecare 63
HomeN 63
HomeNet 63
Homevita 63
homepy see minihomepy

Hong Kong 6, 62, 67, 163, 173, 175, 199, 228
host 45
HOT 167
household robotics 211
HTS 120
humanoid 216
hurry hurry see bballi bballi
high speed 4
HSDPA 99, 169, 180
Hubo 211
Hussein, Saddam 85
Hutchison 163
hypertext 110
Hyundai 63, 138, 144, 178
Hyung, Young Joon 46

I Robot 216
Iceland 6
Iconlab 111, 176, 251
ID card 66
IK Tech 61
Il-Sook, Shin 54
IM 22, 50, 182
iMobicon 225
i-Mode 84, 231
Inavi 143
Informa 157, 176, 250
information society 116
IlikePop 166
Infraware 159
INKA 168
InnoAce 63
Instant Messaging see IM
Institute for the Future 137
Intel 169
intelligent bus stop 68
intelligent car 137
intelligent clothing 63
intelligent home 59
intelligent mirror 71
Intelligent Robot Exhibition 218
intelligent wardrobe 68, 71
interactive 81, 86, 191, 226, 228
Internal Herald Tribune 39, 49

internet 3, 32, 82, 91, 110, 137,
  157, 174, 219, 225, 240, 245
internet cafe
  11, 185
INVIL 131
In Wireless 12
iPhone 64, 79, 155, 160, 251
iPod 8, 23, 29, 155, 159, 238
IPTV 3, 78, 86, 91, 229, 249
Irdeto 126, 139
Ireland 175, 202
Israel 62, 158, 174, 213
Italy 6, 87, 89, 92, 121, 174, 229
IT Ethics 123
IT industry 7, 63, 70, 76, 118, 147, 177, 234
IT Industry Outlook Korea 118
IT Korea Journal 60, 66, 75, 211
Ito, Dr Mitzuko 19, 25, 30
ITU 6, 88, 173, 175, 213, 226
ITU-T 234
iTunes 35, 41, 155, 159, 167
Izen 60

J-Phone 26, 239
Jang, Jae Ho "Moon" 200
Jaokar, Ajit 163, 193
Japan 6, 8, 19, 25, 30, 42, 43, 48, 54, 67, 75, 84, 88, 92, 98, 101, 121, 133, 137, 148, 158, 173, 175, 199, 209, 213, 231, 251
Jaty Electronics 141
jeans (blue jeans) 23
JetAudio 164
Jippii Group 155
journalism 84, 107, 132
Jobs, Steve 160
John Madden Pro Football 191
Journal of Object Technology 241
juke box 45
Jukeon 166
Jung, Yun-Joo 75, 79
Jupiter 212

KAIST 212

karaoke 51, 91, 167
Kart Rider 4, 29, 39, 52, 199, 204
Kasper, Timo 238
KaZaa 166
K-Bank 98, 142
KBS 75, 79, 144
Keio University 19, 25
keitai 30
keyboard (keypad) 110, 187
Kia 138
Kim, Daeho "Showtime" 200
Kim, Rick 225
Kim, Won 241
KIPA 178
KISDI 118
K-merce 98
KMPS 100
Kopomaa, Timo 28
Korea Agency for Digital Opportunity and Promotion 18, 21, 22, 29, 125
Korea e-Sports Association 205
Korea Game Development and Promotion Institute 201
Korea Telecom see KT
Korea Times 17, 180
Korean Internet Security Center 126
KT (also KTF) 63, 68, 90, 98, 129, 138, 141, 144, 158, 176, 184, 234
KWays 142
KyungDong Network 61

laptop 20, 52, 109
LAN 174
landline see wireline
language (Korean) 32
Latin America 85
law enforcement 125
LBS 142
LCD 232
Le Grand, Fedde 164
Lee, Harry 12
Lee, Hyeok-Jae 147

Lee, Ok-Hwa   182
Lee, Woo-Jae   159
learning (and education)   7, 50, 109, 116
legal framework   121
Lego   41
letter   21
Leuwen University   10, 25
Levi's   101
LG   60, 63, 87, 98, 138, 158, 166, 178, 236
LG Electronics   139, 159
LG-Nortel JV   181
LG Prada   161, 252
Liberia   198
Liebhold, Mike   137
lifeline   17
Linked In   49
Lineage   7, 28, 41, 54, 182, 195, 198, 204
Linturi, Risto   59
Lithuania   213
location information   11, 140, 142, 146
London   32
"Lucifer"   200
Lunar Storm   43
Luxembourg   175

M Bank   232
m:Metrics   24, 158
Macintosh   161
MacKinnon, Peter   180
Madden, John   191
Madonna   11
Mahru   212
mainframe computer   50
MainosTV3   239
Malaysia   32, 78, 92, 203
malware   127
Mando Map   143
manufacturing   145
mapping   142
Mars probe   210
Martin, Ricky   158
Massively Multiplayer see MMOG

Mastercard   97, 101
MaxMP3   166
Mayhill   220
McDonald's   26
McKinsey   85
MDA   183
media   7, 80, 191, 225
MediaFlo   231
Melkko, Sara   238
Melon   88, 161, 169
Mercedes Benz   39
Metronerd   51
MIC   68, 71, 79, 117, 145, 211
Microsoft   191
MIDI   162
Miles, Peter   17
minihomepy   34, 46, 49
mini me   41
miniroom   42, 165
Ministry of Information see MIC
Ministry of Sound   164
mischievous   21
MiTV   78
Mixi   43, 48
m:Metrics   64, 140, 197
MMOG (multiplayer gaming)   7, 28, 39, 52, 83, 185, 191, 201
MMORPG   41
MMS   10, 64, 147
Mobile Cyworld   48
mobile internet see internet and also cellphone internet
Mobile Payments World   100
mobile phone see cellphone
Mobile RFID Forum   147
mobile wallet see digital wallet
Mobile Web 2.0   163
Mobile Youth,   24, 27
mobisodes   89
Moneta   98, 232
money see digital wallet
Monopoly   196
"Moon"   200
Moore, Alan   17, 133, 182
Morocco   213
motion sensor   188

Motorola   236
movie see cinema
Mozen   140
MP3   44, 50, 63, 83, 142, 147, 155, 238
MPEG   83
m-payment (see also digital wallet)   4, 98
MSN Messenger   50
Mukebox   166
Multiplayer gaming see MMOG
multimedia messaging (and picture messaging)   10, 64, 147
music   3, 8, 32, 35, 44, 50, 60, 79, 82, 101, 104, 117, 155, 162
Muz   166
MTV   82, 88, 161
multicast   89
MVNO   49, 86, 231
Mylisten   166
MySpace   4, 34, 39, 43, 48, 193
My Sassy Girl   32, 46

Napster   83, 166
narrow band   4
Nate   143, 232
Nateon   50
National Computerization Agency   234
Naver   106
Navermusic   166
Navtech   143
Nbox   141
NCA   234
NC Soft   54, 198
Neowiz   40, 128
Nespot Swing   234
Netherlands   6, 62, 67, 89, 202
Netoro   212
network effect   163
Networked Age   5
New Zealand   21
news   46, 107, 132
Nexon   52, 204

next generation integrated network   234
NFC   149
n-Gage   193
NIDA   24, 25, 40, 65, 83, 105, 141, 146, 156, 159, 183, 197, 229, 240
Nigeria   98
Nike   50, 101, 239
Nintendo   64, 191
Nokia   8, 25, 65, 78, 111, 179, 193, 236, 239
Noh, Jae Wook "Lucifer"   200
Norrathian Platinum   198
Nortel   180
Norway   6, 101, 144
notebook PC (see also laptop)   238
N-Series   194
NTT DoCoMo   101, 143, 176, 231, 239
Nuri Telecom   102

ocean management   122
Ogilvy   21
Oh, Yeon-Ho   115, 132
Ohmy News   84, 107, 115, 132, 182
OIMusic   166
online shopping see digital wallet
Open Gardens   193
OnTimeTek   78
Ottoro   212
Oxford   76, 79

Pac Man   28
pager   144
parents   20, 27
Park, KW   111, 176, 251
parking   144
passport   66
Paypal   100, 227
PBX   242
PC (see also computers, laptops)   51, 59, 110, 137, 160, 187, 217
PC gaming   47, 54
PC bang (see internet cafe)

PDA   52, 116, 143, 181, 187, 238, 250
peak times   87
penetration   6, 10, 75, 85, 229
pervasive computing   129, 173
petrol   97, 144
pets   61, 148
PGR   215
Philippines   31, 98, 101
Philips   82
Piaff Edith   81
picture messaging see MMS
picture uploading   48
pictures (film based, photographs)   26, 48
picture sharing   34
pilot   50
Pizza Hut   27
play   24
Playstation   23, 28, 51, 54, 59, 191
PLC   61
Plus Eight Star   46
pocket TV   75, 238
PocketNavi   143
Pokemon   42
police (see also law enforcement)   8
politics   123
Pong   28
pornography   11
Portugal   62, 121, 175
Porsche   39
Post Guide Robot   215
Post PCs   187
postal banking   100
postal robotics   215, 220
Prada   161, 252
predator   11
prefer   22
President Roh Moo-hyun   123, 133
Presidents of the United States of America   163
preview   90
print   80, 226
privacy   123, 181
professional gaming   200
profile   30, 34

progress   209
Project Gotham Racing   200
projection   187
PS2 see Playstation
pub   41
public bath   11
pull / push   10
Put Your Hands Up (for Detroit)   164
PVR   75, 82, 249

Quake   200

R2 D2   210
radio   81, 85, 87, 163, 169, 226
Radiolinja (see also Elisa)   239
rap   92
Rather, Dan   182
Real Traffic   144
reality TV   82
RealTelecom   144
recording   45, 81
Regal   138
Reigncom   166
remote control   60, 183
Renault   138
replacement cycle   24, 127, 238
restaurant (Korean)   32
retail   106, 145
RF   61
RFID   102, 108, 137, 145
Rheingold, Howard   22
Rihanna   164
ringback tone   44, 50, 156
ringing of phone   32
ringing tone   44, 155, 167
Roadrunner and Wile E Coyote   204
Robo X1   214
robot/robotics   4, 8, 40, 209
Robot Olympiad   218
Robotech   214
Roh, Jun Hyong   115
Rok, Oh Sang   210, 219
Rosen, JJ   160
royalties   45
rules of conduct (cellphone)   184

Sahara 5
Samkyung Hitech 220
sampling 31
Samsung 60, 63, 87, 122, 138, 142, 159, 220, 236, 241, 245
satellite TV 231
Saunalahti 155
Say Club 40
Scandinavia 18
set-top box 231
school 18, 21, 41, 126
search 40, 83
Second Life 4, 7, 34, 39, 100, 182
security 60, 217
Seiko 65
seller beware 105
Seoul 180
Seoul Commutech 63
Seoul Magazine 49, 193
Seven Mass Media 80
Sha Mail 239
Shaw, George Bernard 209
sharing 28
Shinhwa 167
"Showtime" 200
SIM card 98
simulator 50
simulcast 89
silver digitals 107
Sinatra, Frank 81
Singapore 32, 67, 158, 175, 202
sixdegrees.com 46
skateboard 26
SK Communications 30, 34, 225
SK Group 34, 48
SK Telecom 34, 60, 92, 98, 102, 138, 143, 158, 161, 169, 176, 209, 232
Skype 3, 242
SkyPlus 75, 82
Slovakia 198
SM Entertainment 166
Smart 101
smart card 99
Smart Mobs 23

smartphones 11, 143
SMLXL Small Medium Large Xtralarge 133
SMS see text messaging
Snake 193
snacking 79
sneakers 23
soap opera 32, 90, 178
soccer 178
social networking 11, 22, 43, 49
society 116
Soft 143
Softbank 26, 239
Solid Pro (car) 39
Sonata 138
Some Postman 163
Sony 82, 158, 191, 210
Sony BMG 160
SonyEricsson 116, 238
South Africa 98
spam 127
Space Invaders 28
Spain 31, 102
speakerphone 24
SpeedNavi 143
Sri Lanka 198
Ssangyong Motors 145
Star Wars 210
status symbol 27
stolen phone 20
streaming 75
student (see also youth, teenager) 17, 32, 180
submarine 51
SubTV 17
subsidies 127
subway train 97
Sugababes 165
Super Mario Brothers 28
supermarket 97
survey 18, 19, 20, 22, 25, 78, 142, 156, 182
Sutherland, Kiefer 89
Sutherland, Rory 21

Sweden 6, 43, 62, 67, 121, 174, 228
"Sweet" 200
swimming pool 11
Switzerland 6
Syria 198

Taiwan 6, 62, 121, 158, 199
Taejon 245
talk show 82
tamagotchi 52
taxation 120
teenager (see also student, youth) 19, 24, 29, 49
Telecoms Korea 162
Telefonica 31, 102
telematics 137, 142
Telenor 101
Telephia 104, 193, 228
terror 128
test bed 179
Tetris 28
text messaging 7, 11, 17, 25, 60, 62, 84, 123, 147, 180, 234
texting
  and talking 19, 20, 22, 30
  and email 22, 182
  and money payments 98
  and youth 125, 182
  peak 87
  to TV 85, 92, 228
TFT 141
Three 163
Thinkware 140, 143
Tietoyhteiskunnan Synty 28
TiVo 75, 82
TNS 164
Tom & Jerry 204
TomiAhonen Consulting 121, 250
toy 24, 217
traffic info 69, 144, 150
traffic warden 126
Transbot 214
Transformers 214
triple play 90, 232
truetones 156

Tu Media 90, 92, 139, 232
Tubemusic 166
Tunisia 198
TV 3, 8, 50, 60, 75, 126, 169, 191, 200, 226, 249
TV
  advertising 81
  gaming 47
  in car 140
  peak viewing 87
  SMS texting 85, 92, 228
  voting 86
typing 26

Ubistar 138
ubiquitous 5, 104, 129, 173
Ubiquitous Robotic Companion 215
Ubiquitous Dream Hall 68, 71
UI 242
UK (also Britain, English) 6, 10, 19, 24, 32, 35, 42, 43, 48, 54, 67, 75, 86, 88, 104, 121, 175, 179, 183, 193, 197, 228, 242
Ultima Online 54, 194
underground (subway) 97
Universal Music 44, 160
uPostMate 220
Urban Freestyle Soccer 192
URC 215
Uruguay 198
USA (also America) 8, 22, 27, 32, 35, 39, 43, 48, 50, 67, 75, 82, 86, 92, 106, 117, 121, 139, 161, 167, 174, 183, 192, 197, 214, 228
user interface 242
USN Association 145
Uranium Jeans 63
user-generated content 31, 85, 92, 163

VCD 140
VCR 82
VDSL 174, 232
VHS 82
video calling 50
Video Clash 88

videodisk player 140
videogaming see games
video-on-demand 90, 232
video sharing/uploading 34, 48
video streaming 75
Vietnam 198
village 130, 185
Virgin Mobile 86
virtual 3, 7, 28, 34, 39, 49, 83, 195, 200
virtual pet 52, 54, 200
virtuous cycle 10
virus 127
Visa 97, 100
Vive 169
VOD 90, 232
Vodafone 26
voice call vs texting 22, 30
voicemail 157
VoIP 231, 242
Vonage 3, 242
voting on TV 86

waiting tone see ringback tone
wake up 25
Walkman 238, 241
wallet, digital 63, 127
Wall Street Journal 83
WAP 84, 250
Warner Music 160
water 5
Wavaa 166
WCDMA 99, 180, 232
wearable computing 64, 187
web address 26, 110

weather 69
welcoming song 45, 166
Wever 218
WiderThan 156, 161, 169
Wii 64, 191
wiki 84
Williams, Robbie 89, 164
wireless 20
Wireless Watch Japan 251
WiBro 4, 60, 174, 231, 236
WiFi 4, 62, 66, 147, 173, 228, 236
WiMax 4, 231, 236
Wired 197
Vodafone 239
World Baseball Classic 92
world champion gaming 200
World Cup of Soccer 178
World of Warcraft 7, 29, 41, 54, 100, 195, 198, 200
wrapping paper 108
wristwatch 5, 65

Xbox 28, 54, 191
Xroad 141

Y of Convergence 225, 230
Yun, Hae-jong 130
YouGov 19, 20
youth (see also teenager, student) 17, 20, 34, 125, 183, 227
YouTube 10, 34, 41, 93, 163
Yujin 212, 214

ZigBee 102

About the Authors

Tomi T Ahonen is a five-time bestselling author, consultant and motivational speaker based in Hong Kong. He is a guest lecturer at Oxford University's short courses on digital convergence and is the co-host of the popular *"Communities Dominate Podcast"* on the Horizon Channel. Tomi authored the three pioneering books on the business, services and marketing of next generation mobile telecoms as well as the world's first book on the business of digital communities and social networking.

Tomi is known as an evangelist for new technologies who has discussed over 1,000 of his "Pearls" in the public domain, and has delivered keynote addresses at over 200 conferences on six continents and been quoted in over 200 press articles in the *Wall Street Journal, Economist, Business Week, Barrons, Financial Time*, etc. and is often on TV. His columns have appeared in *New Media Knowledge, Market Leader, Ohmy News, Financial Times, Mobile Handset Analyst, Asia-Pacific Connect World, European Communications, IEE Communications Engineer, Telecommunications, Mobile Communications, Total Telecom*, etc. Serving as co-editor of the *Forum Oxford Journal*, Tomi also sits on the Advisory Board of the *Journal of Telecommunications Management*. A founding member of the Wireless Watch, Carnival of the Mobilists and Engagement Alliance, Tomi co-chairs Forum Oxford and blogs at www.CommunitiesDominate.com

Tomi's client list reads like the who's who of high tech, including BT, Ericsson, Hewlett-Packard, Intel, Motorola, Nokia, NTT DoCoMo, Orange, Siemens, T-Mobile, Telenor, TeliaSonera and Vodafone. Tomi has also provided consultancy to major non-technology customers including Aller Group, Bank of Finland, BBC, DHL, Economist Group, HSBC, Kemira, MTV, Ogilvy, Pohjola, Royal Bank of Scotland, Sony BMG, Turner Broadcasting, United Nations Security Council, UPM Kymmene etc. Tomi advises industry associations from the Communications Industry Association of Japan to the Irish Marketing Association and from the Canadian Wireless Telecoms Association to the Infocomm Development Agency of Singapore. For this book Tomi has worked closely with KIPA the Korean IT Promotion Agency and with most Korean industry associations.

Tomi set up his own telecoms, IT and media consultancy in 2001. Before that he was employed by Nokia as its Global Head of 3G Consulting where Tomi also oversaw Nokia's end-user 3G Research Centre. Previously at Nokia he was Nokia's first Segmentation Manager and started his Nokia career with internet gateways. Prior to that he worked in Finland with Elisa and the Finnet Group. His accomplishments include the world's first fixed-mobile service bundle, and setting the world record for taking market share from the incumbent. Prior to that Tomi was Head of Marketing and Sales for New York's first internet service provider, OCSNY. He started his career on Wall Street.

Tomi holds an International Finance MBA (with hons) from St John's University NY and a bachelor's in International Marketing from Clarion University (with hons). His previous books are **Communities Dominate Brands** (2005, with Alan Moore), **3G Marketing** (2004, with Timo Kasper and Sara Melkko), **m-Profits: Making Money with 3G** (2002) and **Services for UMTS** (2002, with Joe Barrett). Tomi is working on his sixth book under the working title of **Mobile as the 7th Mass Media Channel** which is projected for late 2007 release by futuretext. For more see www.tomiahonen.com

Jim O'Reilly is an emerging technology specialist currently working for the Korean ICT International Cooperation Agency (KIICA) as Portfolio and Alliance and Partner Manager for the Europe Middle East and Africa (EMEA) region based out of London. He is an experienced International promoter, speaker and authority on business development Alliances. Working at all levels of Government, commercial, R&D and Media and accurately matching the latest emerging and mass market technologies, intelligence and companies at the cutting edge of Digital IT and Telecoms.

In this role since 2002 he has helped hundreds of leading edge Korean ICT companies and executives with hands on support, strategies, business development and partnering advice in over 20 countries across EMEA. This unique and exciting role had deepened his instinct and appreciation of the complex resources, markets, technology and finances to succeed internationally.

Jim is a passionate technologist, known for his endurance and hunger for developing and marketing new opportunities and still confesses to wake up in the middle of the night with ideas for the next best start-up and venture business plans. Prior to the Alliance manager role for the Korean ICT portfolio, he was co-founder and one of key investors in his own Internet software start-up in Korea. Based near Seoul for nearly three years during the adrenalin rush of the first internet boom in 2000, he created and developed the business plan, models and an International team from 5 countries working and competing within the white hot Korean venture business and Internet explosion. He has built significantly on these first hand experiences of developing technology teams using mature and open source platforms and experienced the ultimate pleasures and merciless realities of being an entrepreneur, investor and leader in the most competitive and dynamic cyber culture in the world.

Jim has a growing reputation for evaluating and developing leading edge technology potential and is regularly consulted for product, solution and company portfolio matching by many of the European Blue chip IT, Telecoms, Broadcasting ,Venture Capital organisations and Governments from developed and emerging countries wishing to share the intelligence, experience and partners required to emulate impressive Korean and ICT growth. He has unique experience covering end to end solutions from chipsets through software and innovative terminals for example sectors such as Broadband, Mobile TV, Mobile WiMAX, IPTV, 3G, VOIP, core networks, Value added services and more.

Prior to his seven years living and working with South Korea, Jim has had various senior management roles with leading edge commercial projects and in technical management leading advanced technology and decision support projects within the ICT Industry for BMW, AT&T, Landrover/Rover, Warwick Manufacturing Group and British Steel.

Jim holds a bachelors of Engineering and Business Studies (with hons) from Sheffield University For more see www.digitalkorea.info and www.jimoreilly.org

Other books by Tomi T Ahonen:

Next book by Tomi T Ahonen

Tomi's next book is on mobile as newest of the mass media, and has a tentative working title of "**Mobile as 7th Mass Media Channel**". The book will examine how the media is changing its perceptions of mobile particularly now in a post iPhone age, and illustrates how Mobile has expanded in capability through the "Eight C's" ie Communication, Consumption, Creation, Charging, Commercials, "Cool", Community, and (remote) Control. The much-anticipated hardcover book is roughly 300 pages in length and is projected for Autumn 2007 release
The book will be published by futuretext

Communities Dominate Brands:
Business & Marketing Challenges for the 21st Century
by Tomi T Ahonen & Alan Moore
foreword by Stephen C Jones Chief Marketing Officer Coca Cola
280 pages, hardcover, futuretext, 2005
second printing 2005
first paperback edition futuretext 2007
ISBN 0-9544327-3-8   **NOW IN PAPERBACK !**

The first business book on social networking and digital communities, from blogging to multiplayer online gaming to mobile phone smart mobs within the context of digital convergence. Covers media and advertising, broadband internet and digital TV as well as the emergence of the power of digital communities, user-generated content and viral marketing. Discusses Generation C (Community Generation), Alpha Users (influencers), the 4 C's. Introduces engagement marketing. Global bestseller 2005, into third printing.

"Invaluable in how power will reside far more with ordinary people than with companies."
Rory Sutherland, Vice Chairman, **OgilvyOne** UK

"An excellent, reassuring book! In 5 years time it will a classic - the new bible for new marketeers."
Dr Axel Alber, Marketing Director, **Masterfoods** Europe

M-Profits
Making Money from 3G Services
By Tomi T Ahonen (360 pages, hardcover, John Wiley & Sons, 2002)
Foreword by Teppo Turkki Strategy Director Elisa Corporation
ISBN 0-470-84775-1

World's first business book on next generation wireless was global bestseller in 2003. Covers revenues, revenue sharing, pricing, profits, mobile services. Icludes 170 service ideas and 50 real services in use around the world. With a clear money focus and covers Money Migration, Hockey Sticks, 5 Ms theory. Includes service creation, revenue sharing, content partnerships, MVNOs.

"One of very few books on 3G services to give comprehensive view of marketing and revenue."
Sophie Ghnassia, UMTS Project Director **Orange/France Telecom.**

"The book is a good read for industry professionals, operators, bankers and analysts."
Voytek Siewierski, Executive Director Global Business Development **NTT DoCoMo**

Other books by Tomi T Ahonen:

3G Marketing
Communities and Strategic Partnerships
By Tomi T Ahonen, Timo Kasper and Sara Melkko
(333 pages, hardcover, John Wiley & Sons, 2004)
Forewords by Mike Short Vice President O2 and Chairman Mobile Data Association; and Jouko Ahvenainen Chairman Xtract Ltd
ISBN 0-470-85100-7
second printing 2004

World's first book on marketing for next generation mobile, and only book on winning in marketshare wars, became fastest-selling telecoms book of all time in 2004. Covers all major issues of marketing including promotion, pricing, sales, branding, distribution channel, and service creation and management, with a telecoms focus including handsets, subsidies, portals, billing. Introduces Alpha Users, Reachability, Murfing, Omega customers, and Reachability.

"Insightful look into how wireless carriers capitalize on their customer data developing targeted marketing."
Jan-Anders Dalenstam, Sr Vice President Business Development **Ericsson**

"Packed with useful and practical techniques to achieve success in the competitive marketplace."
Steven S K Chan, Director Internet Services and Mobile **MobileOne (M1) Singapore**

Services for UMTS
Creating Killer Applications in 3G
Edited by Tomi T Ahonen and Joe Barrett
(373 pages, hardcover, John Wiley & Sons, 2002)
Forword by Alan Hadden Chairman GSM Suppliers Association
ISBN 0471 485500
also translated into Chinese

World's first book on 3G services was world's bestselling 3G/telecoms book in October 2002. Covers 212 service ideas with lots of illustrations, statistics, charts and forecasts. Written by 14 leading 3G experts, for the non-technical reader. Includes the 5 Ms theory, service creation, content partnerships, revenue sharing, marketing and competition in 3G. Covers all major service groupings such as SMS, MMS, m-commerce, video, music, gaming, mobile advertising, infotainment, B2B, B2C, B2E, etc etc

"Insightful discussion into Significant revenue opportunities to bring value to mobile operators."
Dr Stanley Chia, Director Group R&D **Vodafone**

"Explains some of the compelling services the industry will be able to develop and deploy."
Jeff Lawrence, Director of Technology **Intel**

Excerpt from Tomi T Ahonen previous book
Communities Dominate Brands with Alan Moore
(published by futuretext)
2005 first hardcover edition
2007 paperback edition
ISBN: 0-9544327-3-8

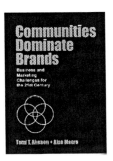

"This book is invaluable in predicting how the power to make and break brands will reside far more with ordinary people than with companies."
    Rory Sutherland, Vice Chairman & Creative Director, **OgilvyOne** UK

Foreword

A few weeks ago I was visiting The Harvard Business School as a guest lecturer and during a break was sitting in Spangler Hall reviewing Tomi Ahonen's and Alan Moore's powerful new book, *Communities Dominate Brands* when I looked up at the dozens of groups chatting away and wondered "What the hell are these students going to do with this insight and opportunity. Should I even share it with them at the risk of blowing their finely tuned minds? Or maybe they'll dive into their first post grad job determined to implement such bold directions." So I shared some of Moore and Ahonen's thoughts later in the lecture hall. There wasn't one student who didn't believe that they were not a member of a virtual connected community. Soon we'll see what they do with it as leaders.

Five years ago, January 2000, I became Chief Marketing Officer of The Coca-Cola Company at the height of the dot com craze and fresh out of Japan where I had lived and worked for the previous six years. When I arrived in Japan there wasn't a cell phone to be found but a year later DoCoMo introduced them for the price of one yen (and a hefty monthly fee) and a few months later they had sixty percent penetration. Months later the Georgia Coffee team and DoCoMo introduced the 'ring tone down load' concept using the famous Georgia Coffee jingle that took the country by storm. At about the same time internet use soared from a few after hours office workers to a national phenomenon. We played outside the traditional marketing box with some entry level internet marketing promotions which unintentionally started a dialogue with six million consumers! Wireless cellular technology and the internet forever changed the consumer and our approach to marketing in Japan. And it reshaped my own mental model of how to engage in a relationship with consumers.

As the new CMO facing an unprecedented period of change, I asked Anne Chambers, our resident enthusiast to search for best in class examples of the how others were engaging in what I had experienced in Japan, and surprisingly we found them either unaware or so fearful that they would lose control of the brand message that many had actually banned the use of the internet as a marketing tool. Only a few were experimenting but they were operating in the world of wireless technology. I had read some of Alan's material and found a sense of excitement that validated my own intuition that the virtual community I saw developing in Japan couldn't be stopped, mustn't be stopped. One of my closest advisors, Nick Donatiello, CEO of Odyssey, who was an early pioneer of internet research and a brilliant strategist, gave the best advice possible. "Just jump into the dialogue. Let's see where it takes us."

Available now also in Paperback, Tomi T Ahonen's previous book **Communities Dominate Brands**

It is difficult to put a lens on a developing social trend moving as fast as 'connected communities' but Alan and Tomi have done that. Together they have made a rare and important breakthrough insight, have developed a credible hypothesis and backed it up with validated supporting points. This is not radical misinformed extremist hype. This work is an accurate description of the issue, the opportunity and the crisis confronting marketers if they don't cut loose the shackles of the traditional advertising agency and TV network model and explore the world of possibilities recommended by this book.

Move quickly but act thoughtfully, even slowly. You want to implement this without sending your organization into a tail spin. The traditional marketing company that wastes its investments solely on TV advertising is underpinned by bureaucratic values of safety, efficiency and control. The marketing group that embraces these insights and moves forward to implement them is underpinned by interdependent values of sharing, listening, equity rights, global harmony and synergy. That's a big leap. Tomi and Alan are not proposing a process tweak but a mindset shift, one that requires an evolution of values and a transfusion of talent. Their thinking is visionary. To succeed with his model you need to line up your organizations vision with the prerequisite values and talent. Do it. But do it thoughtfully so that everyone understands and believes the plot first. You'll succeed with lightning speed if you do. You risk crashing and burning if you don't.

I am a believer in Alan and Tomi's insight and forecast. The consumer and their connected communities, selecting the products and brands that are engaged in the most relevant dialogue with them, is the center of any modern and sustainable marketing model. Wireless technology has enabled the consumer to review and reject much of the one way messaging they receive and resort the dialogue that's relevant to fit the way they live. Experiencing a Coke or interacting with an enthusiastic Coke employee on line or in person has always been far more motivating than 30 seconds of anthemic brand worshipping. It's not that TV and radio programs are irrelevant. It's the lack of ability to develop a relationship with an ad that makes the medium a less viable marketing tool.

Books on business and marketing are launched weekly. Most are weak adaptations of other people's thoughts. Some authors like Sergio Zyman, Seth Godin, Scott Bedbury, and Marc Gobe, have made bold and meaningful interpretations of contemporary opportunities and helped me to clarify a new advanced perspective on how to be a more successful marketer. Tomi and Alan have done that and with **Communities Dominate Brands** will end up shaping our thinking and approach for some time.

Stephen C Jones

Published reviews on **Communities Dominate Brands**:

"This is an eye-opener with a key message essential for all consumer centred enterprises. An excellent, reassuring book! In 5 years time it will be called a classic - the new bible for new marketeers."

Dr Axel Alber, Marketing Director, **Masterfoods** Europe

"This book provides a comprehensive understanding as to why business, media and customers will never be the same again; where interrupting audiences and one-way flows of marketing communications are things of the past."

Rishad Tobaccowala, Chief Innovation Officer, **Publicis Groupe Media** USA

Excerpt from Tomi T Ahonen & Alan Moore 2005 bestseller *Communities Dominate Brands*

> "When written in Chinese the word crisis is composed of two characters.
> One represents danger and the other represents opportunity"
>
> John F Kennedy

I - Introduction
On the Road to Engagement

If the last 10 years have caused disruption in your business, the next 10 years will cause much more so. Not driven by a controlled introduction of new technologies, but by an uncontrolled adoption of new, radical, unpredictable and even "unfair" methods by an emerging new element in consumption – the *digitally empowered community*. Digitalisation and the falling costline of technology have ripped through our business and social fabrics over the last five years, across all industries, across all countries, altering our economic and social landscapes forever. Yet what we have witnessed so far is only the beginning of a more profound, *seismic shift* in the very foundations of how business is conducted. We can imagine and do things which were just not possible a few years ago. Life-threatening or life-enhancing? This is what this book is about; the Red pill or the Blue pill? Which one are you going to take? One thing is for sure: the structured order of our familiar industrial age has come to an end and it's dying as days do, gasping for every last ray of light.

All the rules are changing

Now, the Age of Connectedness and its newly active communities are altering the way all businesses will market, promote and sell their goods and services. The very first cases are emerging simultaneously around the world, and they clearly give an answer to what the marketing industry has expressed for several years already. Traditional methods of marketing, advertising and branding are increasingly ineffective. Something new is happening, only we could not put our fingers on it. Not yet. Not until this book.

Phillip Evans and Thomas S. Wurster state in their book *Blown to Bits* that digitalisation is "deconstructing" traditional industries such as home electronics, business, broadcast, retailing and banking, while at the same time creating new commercial opportunities such as Google and eBay, the low-cost airline industry, online banking, or Closed Audience Networks. These are trends echoed time and again by experts analysing the individual phenomena such as digital convergence and disruptive technologies.

Today, central to young people's discussions are music players, mobile phones, enhanced bluetooth technology, digital cameras, robots, plasma screens and laptops. From Tokyo to New York this is the digital generation, brought up on generating their own content or consuming the content they want, digitally. So profound is their impact that the *Financial Times* writes that companies like Time Warner, Sony and Walt Disney are being forced to rethink their business models. Yes, we know the young are digital, but that is only half the story. There was a digital generation a decade ago, with Playstations, personal computers, digital calculators, portable CD players and Nintendo. *Digital* is *not* what is different this time; digital is not the key. The change is something deeper and more profound.

The new digital economics has removed the need to decide between whether one has richness or reach. Today, you can get both. This changes essentially everything. It changes the way customers can access information and changes the way they use it. It changes the way business can communicate with their customers and it also changes how a business might go to market. It changes the linking between channels, that link businesses, customers, suppliers and employees. It offers opportunity and it offers your once helpless competitors the chance to radically rethink their business strategies and attack vital parts of your business model. New and hungry players are taking every opportunity to enter the value chain, hoping to disintermediate you and your brand promise.

We are still observing the very beginning of this, business guru Gary Hamel says: "The least appreciated effects of digitisation is the fragmentation of customer attention. Customers become harder to find and more difficult to keep."

From a Networked Age to the Connected Age

Much has been written about the digital age being the networked age; that we plug into and out of the network and get considerable benefits from being connected to the network. The network age was a good term for the 1990s, as it did describe how we as humans approached "the network" – ie, the internet. We logged on, we accessed our email and we surfed seeking information. Much of what most readers will consider the digital world and digital convergence will consist of that networked model.

The first decade of the 21st century starts mankind's next evolution in delving deeper into the information age. We move beyond the networked age into the "connected age". Typical of the connected age is that we no longer have to physically log on and log off. We are not tied to any single physical place to find our connection. It is not the office or the home where we have our connection, and we do not have to connect at a hotspot. We are always connected and we can instantly access the network. We can be reached at any time, and typical of the connected age, we start to manage our connectedness when we deliberately disconnect ourselves for personal reasons.

The single most visible change from entering the Connected Age, is that we suddenly have permanent access to our peers, our friends, our colleagues and family members. We can start to live with a "lifeline" to those we trust. Our communities, which previously only existed at given points in time, now become ever-present. We are no longer alone. In the Connected Age modern people are able to draw on the community for assistance, information and support. We learn to search, share and interact in a new way.

In the Connected Age people will have public and private – and semi-private – personas, which coexist in the network and are connected independently. We may want to keep our public persona connected only during office hours. We may want our private connection always on, but always with the ringing sound turned off with all personal contacts knowing to use SMS text messaging to reach us. And we might connect and disconnect with our semi-private persona; for example, relating to our hobby or passion, be it football, car racing or opera.

In the Connected Age the intelligence and ability to customise our gadgets becomes ever more powerful. We will find that most of the novelty of surfing the net has worn off, and we rarely surf just for the sheer joy of discovering new websites. Our devices learn to adapt to our whims and preferences and quickly help us navigate to the sources of the information, entertainment and utility that we seek from the web. That kind of content and related applications will be increasingly consumed on mobile or cellular devices. When we use our cellular devices we will usually be in hurried states and need access fast. For that speed we are willing to pay something, and that payment in turn helps keep the content at our favoured sites current and valuable.

But, with the greatest of threats comes also the greatest of opportunities. So, how do you navigate this newly converged world and what are the strategies that will enable you to do this successfully?

Brands in paralysis

We will show in the book how traditional advertising, marketing and branding are in crisis and how traditional marketing communications are becoming bottlenecks for growth. Brilliant marketing minds of a generation have been harnessed to deliver marketing from its despair, yet none have shown a sustainable method or tool for the marketing industry to deliver. All experts agree there is a problem, none of the solutions have been found to work. Foster and Kaplan state in a recent book entitled *Creative Destruction*:

> *Corporations are built on the assumption of continuity; their focus is on operations. Capital markets are built on the assumption of discontinuity; their focus is on creation and destruction. The data present a clear warning; unless companies open up their decision-making processes, relax conventional notions of control, and change at the scale and pace of the market, their performances will be drawn into an entropic slide into mediocrity.*
>
> Foster & Kaplan, *Creative Destruction*, Currency 2001

Technology changes the way we interact, the way we shop and consume. It creates new opportunities and destroys businesses that are unable to adapt to a sudden discontinuity with our past. We are moving from a production-driven to a consumption-led economy, where the nature of exchange is different, and this difference is exacerbated by the forces of digitalisation: the internet, e-commerce and the mobile phone.

Our recent history has been deeply affected by the increased speed of technological development plus the convergence and proliferation of the audio-visual, mobile, IT, and personal computing industries, increased internet and bandwidth penetration, and media choice. These developments have impacted on the businesses and the marketing community. As a result of these developments, business itself is faced with a tougher job when innovation and flexibility are the markers for competition, rather than efficiency being the fundamental driver of value.

We see creative destruction from disruptive effects to digitalisation to disintermediation by network effects in industry after industry. The music business, the movie business, television broadcasting, banking, the airline industry and travel, publishing, retail, utilities, government etc. The effects are seen everywhere. Michael Nutley, the editor of the *New Media Age,* says: "Industries which try to dictate how their customers should transact with them are taking a huge risk in the digital age." As a consequence of the uptake of new technology and the way customers' habits are changing there is a lot of "creative destruction" happening across industries.

Enter the community

A key development that has been monitored and noted in numerous instances is the emergence of digitally connected communities. Mostly these have been seen in isolation, as a new "market space" opportunity to harness and harvest, to exploit if you will; to make money from. Communities like eBay's online auctions and shopping, or communities like online dating, music and movie file sharing, friend-finders and job recruitment, etc. These are significant developments by themselves. But they are the earliest visible symptoms of a massive development in human behaviour and change in society. Many more powerful and personal communities are also forming, and these are not limited by the shortcomings of the fixed internet.

Communities that use mobile phones to share, influence, connect and participate are spontaneously emerging in all countries in all areas of interest, from car shopping to birdwatching to anti-government revolt. Rather than the armchair networkers of fixed internet communities, those who connect on mobile phone communities are young, mobile and active. They can suddenly swarm and appear by literally the thousands on a moment's notice. These "smart mobs" as Howard Rheingold wrote in his book of the same title in 2003, can become activists either for or against any authority, politician, company, product or service.

With the first actual evidence of community activities on behalf of some product, services and companies, and of evidence of communities also against companies, we can formulate our thesis: that communities are the counterbalance to brand dominance in the 21st century. By examining what word of mouth, when enhanced by the powerful digital echoes of mobile phone communities and other networked communications such as IM Instant Messaging, email, and increasingly blogging, can do, we see that communities have been an undefined or underestimated barrier to recent marketing success. Furthermore, we establish that by harnessing community power and working with them, a modern marketer can succeed in delivering remarkably positive marketing effects to the intended target audience.

In our book we proceed logically by looking at the disruptive trends of technology, chance, digitalisation, disruption, convergence, and societal changes, then how businesses are changing. We examine how branding, marketing and advertising is in crisis. We establish the basis for understanding communities, both in the Networked Age of the past decade and the new Connected Age. We analyse the emerging new type of consumer. Still mostly under the age of 25, this Generation-C is the *Community* generation and we show how dramatically different it is from older generations, and how intuitively it already uses community power to its own gain. We then show how communities and brands interact, and why communities dominate brands. And finally, we show how businesses can thrive in this new marketing environment: they need to evolve from interruptive advertising to engagement marketing.

Bloggers, gamers or Gen-C?

We devote three chapters throughout the book to discuss in more detail the community behaviour of three groups of digitally-aware societies. In the virtual chapter we discuss videogamers. In the blogging chapter we discuss fixed internet bloggers of today, and the likely emerging mobile bloggers of the near tomorrow. And in the Gen-C chapter we discuss the young cellular-phone connected smart mobs. These are not the only digitally connected societies. There are countless more using the digital communication tools of choice most suitable for each community. We discuss gamers, bloggers and Gen-C because these three groups seem to have evolved furthest into discovering community power now, in 2005, and from a business point of view. We must emphasise that digital communities are inherently self-improving; they will evolve dramatically over the next few years. It is safe to assume that all of the power of the exceptionally successful community action that we describe in this book today will be totally commonplace with all communities a year or two down the line. Make no mistake about it: no matter what your business, your customers will behave like these communities.

Excerpt from Tomi T Ahonen & Alan Moore 2005 bestseller *Communities Dominate Brands*

Why us

We reveal a change to how all businesses need to interact, from the branding and advertising to all core marketing activities. The change is enormous, the biggest single change in business for the past 100 years. Why suddenly is it that the two of us would discover such a profound change?

We have been lucky to be involved in several strategic marketing projects in the earliest markets where this phenomenon has started to happen – the countries where mobile phone penetrations reached young teenagers first; Scandinavia in general and Finland in particular. We have also been lucky to work with those leading companies in those markets whose very core competence and deepest research happens to hit this area where our new Generation-C, with C for Community, has just emerged. We have supported companies that have specifically worked to understand Scandinavian youth and how it interacts with mobile phones. Our customers have included Nokia, Ericsson, TeliaSonera, Elisa, and numerous media companies, as well as support organisations with the deepest customer insights; such as specialist Xtract the user profiling company, or Fjord Networks, the User Interface company.

Even with exposure to the very earliest user patterns, it took us collaborative thinking and analysis and our own research to finally develop the theory that combines the counterbalancing forces of brands and communities. However, with every emerging new finding our conclusions become more firm. The facts solidly support our hypothesis and we can now already claim it to be true, the balance is in favour of the communities: communities dominate brands.

An American angle

Some of the issues in this book, particularly the early parts of digital convergence and disruptive technologies, are very familiar to American readers. The ideas of blogging and virtual environments are also not alien. And, as we all know, Americans tend to be world leaders in innovation in marketing, advertising and branding.

American readers should pay particularly close attention to the issues of the cellular phone, the Connected Age and Generation-C. Because of the early successes of various digital, internet and wireless data solutions, American businesses may be blinded to the "big picture" – the much more dramatic shift happening to communities activated by the cellular phone. It is not a "rival" technology to co-exist with email, e-commerce, IM Instant Messaging etc. No, cellular phone based communication is a *total cannibalisation* of the digital space.

It is very easy to become impressed with the dramatic growth rates of the various digital delivery technologies from broadband internet to digital radio. Yet these all pale in comparison with the pervasiveness and power of the cellular phone. Do not become distracted, the analogy is not railroads vs airplanes – both of which still co-exist in the 21st century. The appropriate analogy is steam-powered cars vs gasoline-powered cars. In 1890 over 90% of all motor driven vehicles were steam-powered. By 1920 less than 1% were so. Understand the shift from the Networked Age to the Connected Age. The future of every business depends on capturing the soul of Generation-C, sooner or later.

It's all about dealing with change

In the final analysis this book is about change. The early parts of the book echo, repeat, summarise and explain the related relevance of several known current trends in changes that affect business. These changes, while significant and disruptive and even frightening for some players, are known. Most organisations should have processes and plans in place to prepare for those changes, to capitalise on any opportunities and prepare for threats.

The big, monumental idea in this book is that marketing will have to totally change. When we say companies must move from interruptive advertising to engagement marketing, we do not mean that it is one campaign. It is the whole business, a change to the very fundamental way a company's marketing is planned and executed. Engagement marketing, when fully embraced, is a radical thought and bringing about that kind of change will invoke resistance from all parts of the organisation, from the inside – established marketing managers, advertising and product management personnel and branding executives – to the outside partners and suppliers such as the advertising agencies and PR agencies etc.

The benefit to the bold is the rewards of being first to the future. The trends we have identified are proven to be inevitable. All significant authorities in each of the related fields is as near unanimous to the major trends as can be expected. But nobody had shown a way out of that quagmire until now. We show the way: it is engagement marketing, and we give a few early examples of real holistic changes to marketing activities that have been designed to embrace engagement marketing. For the reader of this book, if you find the individual trends to be self-evident, then the final conclusion will also have to be so.

We urge all professionals in marketing to stop projects involved in mindless propaganda of interruption advertising and branding. We urge them to learn how to introduce engagement marketing and harness the power of communities. If it is not you, it will be one of your competitors. That is the final certainty proven in the digital age. Any new good idea will not remain secret for long. When a book like this pushes engagement marketing, if you are not the company introducing it to your industry, your most dangerous competitor is the one to do it.

The future of marketing will not be built around hits, cut through, or attempting to bring as many people to a media channel at a single fixed point in time. It is like applying military strategy from the medieval ages to the battlefields of asymmetrical warfare of today. It is rather by creating compelling reasons for people to engage with your brand and to aggregate audiences over time, when they can come to you when they are ready. This is based upon insights into the fundamentals of human nature, combined with the push and pull capabilities of technology and integrated media channels. It is also potentially a lot cheaper.

Dominate!

You know your marketing, branding and advertising is not working. You know your industry is facing threats if not from actual digital rivals, then your processes face digital disruption. This book reveals the enormous power of the community, if it is not already countering your marketing, it very soon will be. We provide a thoroughly researched manual on how to succeed in the future, illustrated with 13 real business case studies from around the world. If you adopt engagement marketing methods you can succeed in the Connected Age. You can work with communities. In fact you can dominate. But to achieve that you must be bold and change. Not a small change, a huge change. It takes courage. With that we are reminded that it is not the biggest or the brightest that survive change. As Charles Darwin wrote in *Origin of the Species*: "It is not the strongest that will survive, nor is it the most intelligent; but the one most adaptive to change."

Excerpt from Tomi T Ahonen & Alan Moore 2005 bestseller *Communities Dominate Brands*

> "Its easy to get stuck in the past when you try to make a good thing last."
> Neil Young

II - Society Changing
Discontinuity, does it hurt?

Culture has collapsed into the marketplace and vice versa. Our concepts of institutions and industry sectors once living in splendid isolation, ring-fenced from each other today seem antiquated. Pachter and Landry in their book Culture at the Crossroads believe that the spirit of our age is one in which the sciences and economics question and challenge the notion of fixed categories, perceived oppositions, and impermeable boundaries. This means our culture is closer today to notions of flexibility, fluidity, portability, permeability, transparency, interactivity, and engagement. We will start the analysis of the trends affecting our lives with the general ones, and then proceed to the more specific. Therefore this chapter will examine how society as a whole is changing, with institutions and our lives overall facing change.

A VALUES CHANGING

For most of humankind's history the basic values have been consistent, rather than constant, changing only gradually over time. Furthermore, the geographical distance between societies had kept developments distinctly out of synch. Famously Nepal had not had television until the 1990s, and most Nepalese held totally different values from most of the rest of the world. But with the introduction of TV, they suddenly adopted very Western, brand-oriented, spending-oriented status symbol- laden values. While telegraphs and telephones did technically connect "everybody to everybody" all through the last century, there was little practical need to call up a random person in Bolivia, or Burma or Belgium just to talk to them. It was not until the internet in the 1990s that suddenly it became quite common to exchange thoughts with total strangers who could be in any random foreign country.

Starting with radio and more strongly via TV the world started to harmonise its values. It wasn't until the internet that communication in a "world community" became truly viable. This society of the new information age holds surprisingly different values from older generations, and also remarkably consistent values between themselves, globally.

Death of intimacy

Our privacy is being eroded from several different directions. We reveal more and more of our preferences to faceless institutions that collect data on our behaviour. In London you are on closed circuit TV for the police in practically every public place including streets, shopping areas etc. Border controls are becoming significantly intrusive; now the USA already takes fingerprints and digital photographs of everybody who enters the country.

There is an ever-increasing concern over the erosion of privacy. We see individual intrusions everywhere on a daily basis, starting with spam via the internet. The writer Martin Jacques identified three important trends that change society in an article entitled "The death of intimacy" in The Guardian. The three trends are 1) the rise of the individual, 2) the spread of the market into all aspects of society and 3) the rise of communication technologies, notably the mobile phone and the internet.

Jacques has made observations which are significant and thought-provoking, and there are important lessons that apply to businesses and brands. If your business wants to offer added-value then you have an increasing responsibility not to be invasive or intrusive. Recent research, such as the British Chartered Institute of Marketing report You talkin' to me?, (January 2004) and the Yankelovich Survey Report (April 2004) both confirm that people feel that brands are more intrusive today than before. These are but two examples of an avalanche of similar reports.

Self-actualised people

Maslow's Hierarchy of Needs suggested that as people grow content with more primitive needs such as food and shelter, they can move up towards the top of the pyramid, to ultimately seek self-actualisation. In 2001 SMLXL conducted research among teenagers in Germany, Spain, Italy and France. The findings reveal that young people today are remarkably self-actualised. In fact a direct statement from an interview sums up the findings: "The other side of fear is freedom." This generation is the first to grow up without a significant fear of disease and hunger (1910s, 1920s), of poverty (1930s), of war (1940s), or of nuclear war (1950s – 1980s). The findings were remarkably (Continued...)

Published reviews on *Communities Dominate Brands*:

"Although wary of another book claiming that the world has forever changed, I have been won over by this deeply impressive book. Packed full of statistics, examples and case studies, the arguments are well supported and persuasive. The authors provide a comprehensive exploration of this emerging topic which is presently unrivalled. Thought-provoking and practical, you will be hard pressed to find a more challenging marketing book this year."
UK Chartered Institute of Marketing (CIM)
(official book review on behalf of CIM)

"The authors vividly illustrate the rapidly growing power of digital communities with examples of real cases where companies have achieved considerable business success by engaging customers."
Harry Drnec, Managing Director **Red Bull** UK

"All other books on marketing pale before this book on the 21st century world. This is the world of my children rather than my parents. A must read. Written with verve and excitement. I can see neurons humming. I am assigning it to my classes at University of California as a required text."
Professor Richard Ross, **University of California Santa Barbara**, USA

"An absolute cast-iron must-read. If you have anything to do with marketing, mobile, advertising or the media this is essential reading. It's a wake-up call for anyone who thinks today is just like yesterday, just a little bit faster. Read it and you WILL want to change the way your business functions."
David Cushman Projects Editor and Engagment Evangelist, **emap** UK

"Great book with a key message for our business about engagement. Consumers want a relationship with companies and they want - indeed expect - to be treated with care and respect. This book has changed the way we look at our business and our relationship with our customers. Good research, background and case study examples including where it can all go wrong. Very Good Book."
Rob Castle, Managing Director, **Korg** UK

Excerpt from Tomi T Ahonen & Alan Moore 2005 bestseller *Communities Dominate Brands*

ABBREVIATED TABLE OF CONTENTS:
Communities Dominate Brands

Foreword by Stephen C Jones Chief Marketing Officer Coca Cola
Acknowledgements
Chapter 1 - Introduction
Chapter 2 - Society Changing: Discontinuity, does it hurt?
Automobiles, search, locks, shoes, TV, text messaging
Case - Transistor Project
Chapter 3 - Business entities transforming: From dinosaur to puma
TV, music, airlines, sports, newspapers, telecoms
Case - Apple i-Tunes
Chapter 4 - Services and products fragmenting: A street brawl with no rules
Cameras, music, movies, newsmedia, airlines, cosmetics
Case - The Guinness Visitor Centre Dublin
Chapter 5 - The emerging virtual economy: Magic kingdoms for you and me
Virtual worlds, video gaming, search, ratings, e-Commerce
Case - Habbo Hotel
Chapter 6 - Delivery channels splintering: Battle of the channels
Newsagents, department stores, video rental, internet, TV
Case - SMS to TV Chat
Chapter 7 - Blogging: Everyone a journalist
Automobiles, newsmedia, sports, festivals, politics
Case - Kryptonite
Chapter 8 - Customers changing: Brand polygamists or worse
Education, texting, TV, books, gaming, art, restaurants
Case - Ohmy News
Chapter 9 - Generation-C: The Connected Community
Text messaging, fashion, gaming, politics, dating, telecoms
Case - Star Text
Chapter 10 - Advertising in crisis: If we say it just one more time
Television, advertising, music, movies
Case - Tango soft drinks
Chapter 11 - Branding losing its power: Is anyone listening?
Television, soft drinks, music, retail, movies, automobiles
Case - Thomas Cook TV
Chapter 12 - Emergence of the Community: Trust your friend
Music, telecoms, politics, dating, shoes
Case - Howies
Chapter 13 - Communities dominate brands: Mars and Venus
Newsmedia, automobiles, tourist guides, IT, wineries
Case - Twins Mobile Music Service
Chapter 14 - From disruption to engagement: Capitalising on Community
TV, e-business, aerospace, electronics, music
Case - Orange bicycles
Chapter 15 - What next?
APPENDIX: Abbreviations, bibliography, websites, blogsites, index, about the authors

NOTE: Tomi's next book will be on **Mobile as 7th Mass Media Channel**

Excerpt from futuretext book *Mobile Web 2.0* by Ajit Jaokar and Tony Fish

Excerpt from futuretext book
Mobile Web 2.0
by Ajit Jaokar & Tony Fish
2006
ISBN : 0-9544327-6-2

Chapter 1
Mobile content and the changing balance of power

Synopsis
Content created on the mobile device will change the balance of power in the media industry. The mobile device is ideally poised to capture user generated content 'at the point of inspiration' – making it the main driver behind Web 2.0. This new role is affecting the role of other 'screens' in our life (such as the PC and the TV) and is triggering a change in the balance of power in the content creation industry.

To explain these concepts, we have to understand a range of ideas that are at the boundary of technology and user behaviour.

In the following sections, we shall first discuss the importance and significance of user generated content and then discuss the impact of mobile devices on the value chain.

The six screens of life and the existing flow of content

For the most part, we are consumers of content. In our daily lives, we consume professionally created, produced and edited content from traditional and new media providers on our "six screens of life"

These "six screens of life" are divided into two broad categories, big screens and small screens, each with three sub groups

The 'BIG' screens of life
Cinema (shared with other members of the public):
TV (shared privately within our homes)
PC (personal or shared use)
The 'small' screens of life
Fixed/Portable Players (fixed devices in things that move such as cars, planes, etc)
Information screens e.g. iPod, radio.
The mobile device, an individual and personalised handheld device

Both for Big and Small screens, the user has traditionally been a passive receiver of content (i.e. content has been broadcast to the user) OR the user has been seen as a member of a carefully controlled and managed audience (for example voting) – but not as a primary creator of content

Copyright © futuretext Ltd 2007   Tomi T Ahonen & Jim O'Reilly

For instance:
- The TV needs users to consume(view)
- A web site needs users to consume/interact in most cases
- Cinema needs users to consume

Reflecting the above trend, most of the content on the Mobile to date, has also been the 're-presentation' and 're-production' of existing material delivered to the mobile screen. This chapter will discuss how the mobile device is changing from being a primary consumer to a major creator of content.

The supply side

Some content is created by a professional media companies (for example: movies, music), and other content by the individual (for example: blogs, wikis). Eventually, the content is consumed in a private or a shared fashion.

The platform

Content created from the supply sources is passed through a platform. The platform has two components of service: - the physical components and the enabling components.

Examples of the physical components are:
- The transport network: which provides a means to carry the data from production to the point of access
- The access layer: that allows the user to connect and receive the content
- Storage: the component that holds the content until it can be consumed
- The database: that makes content available for searching

Examples of the enablers are:
- Search – the engine that drives the ability to search
- Identity – who people are
- Sync – synchronisation to ensure quality of service
- Navigation – enabling ease of use
- Payment services - to ensure that economics work
- Directory – ability to find
- Rating – what is good and what is not good
- Location – A mapping of location and content
- AdSense – ability to serve ads within content
- Security – protection

The platform is built by companies who want to take content and deliver it to customers. Each company will offer different physical and enabling components depending on their business models.

Demand

Demand is classified into a number of services, each broken into shared and personal modes of consumption. These services include:

- Communication: Shared services such as video and pictures and personal such as IM, SMS and email
- Information services: Information is a matrix of immediate or historic delivered in either a rich or plain format. Examples of information services could include information delivered on the TV in a rich media format and the same information would be delivered on the mobile browser in a text format.
- Video, Image and Music: Like information, this group is a matrix, depending on sit forward or sit back, and adjustments for the quality of the service needed. Examples would be CD quality vs. MPEG3; depends on listening in the living room with $4,000 HiFi vs. an iPod on the train.
- Games: Games and gaming is a complex area and is partly driven by the platforms but also by titles. However, like other demand services, it splits into personal and shared for playing.

Channels

The channels of delivery are broadly broadcast, session IP and narrowcast.

Broadcast represents TV and other broadcast channels. Session/ IP represents the Internet and narrowcast represents the mobile device. Unlike Broadcast, narrowcast has fundamental limitations due to the physics of the device itself, battery life and radio propagation.

The model (supply – platform – demand – channel and consumption) ends with the content created being consumed on a variety of screens on a shared or a personal basis.

Consumption

Consumption is based on the six screens of life. To recap, these are:

The 'Big' screens of life
- Cinema (shared with other members of the public):
- TV (shared privately within our homes)
- PC (personal or shared use)

The 'small' screens of life
- Fixed/Portable Players (fixed devices in things that move cars, planes, etc)
- Information screens e.g. iPod, radio.
- Mobile-communicator, an individual and personalised handheld device

These screens could be classified into: Fixed, carried and carry screens

Fixed: Fixed screens are physically confined to a space for example the TV, PC.

Carried: A carried device is a device which could be on the move BUT is not personal. In flight entertainment is an example of a carried device.

Carry devices: Carry devices are portable, lightweight and free of wires, the prime example being a mobile phone

Further, the devices could be classified into:
a) Shared : where more than one person could view the screen and
b) Private : where one screen corresponds to one viewer

We make two observations with respect to consumption patterns:
a) Consumption is becoming increasingly private especially with mobile(carried) devices and
b) Private consumption needs an understanding of the context and the content. Merely transforming content from the shared, bigger screens to private, smaller screens is not enough. We explore the idea of context in subsequent sections in detail.

Specifically, in relation to mobile services, we note that existing mobile services are relatively simple and don't take into account the content or the context

Like the Internet, mobility is all about communications. The end user is the 'killer application' providing both the content and the context to services such as voice, e-mail, txt and IM.

When we initiate a conversation (contact) we take into account the context of the message – for example, who we are communicating with (Partner, Work Colleague, Friend, Financial Advisor, Utility Service Provider, Customer, etc). This context also drives the content of the message (what we are actually
saying). We also take into consideration other parameters, such as the time of day, time we have, the level of urgency, emotional state, location, the bill payer, etc.

So far, the network operators, fixed and mobile, have been successful in provisioning these en-mass communications services. These services lend themselves to provisioning on a large scale and require broad based (un-targeted) marketing to their customer base. They are not complex and inherently price driven.

As communications markets mature (both in penetration and in reach), new value will be derived from applications which lower costs for the end user, leverage the contextual and situational aspects and/or exploit disruptive technologies

Content is King – unless the creator changes

The content value chain as depicted earlier, is oriented towards broadcast content. True, there are some user generated components in that chain – such as blogs etc, but the value chain itself is designed for pre packaged broadcast content.

Hence, as we have seen previously, the prevailing industry notion amongst members of the broadcast content industry is: 'Content is King', which runs contradictory to our thinking outlined previously in this book.

However, the notion of '(Broadcast) Content is King' changes dramatically when user generated content becomes the dominant content type.

In this section, we introduce three ideas:

Excerpt from futuretext book *Mobile Web 2.0* by Ajit Jaokar and Tony Fish

a) Content creation is triggered by events
b) User generated content is increasingly being consumed by the community and
c) The community could take on some of the functions of the editor.

To understand these concepts, let us consider the basic nature of content itself.

Content is created due to the occurrence of an event.

No matter how it is triggered, content falls into four value brackets:
• Unique content: such as a film epic whose value increases or decreases depending on marketing and fashion (trends). Unique content could also include a piece of art work or innovation, whose value increases with each passing year
• A body of work: such as a database that could be used for statistical analysis.
• New and News: A new story or a re-release of something that has come back into fashion. In just about all cases, news value falls off within 24 hours for there is something newer. Its value is short and time limited.
• Brand: in rare cases, an event will become a brand, such as Live81, which, if it reaches unique status, produces increasing value over time.

There are a wide number of commercial techniques that can be employed for the exploitation of content generated by these events; which can include the sale or licensing of rights for use.

The content thus produced was originally consumed only in the mass market media.

Now, that content is increasingly being consumed by the 'community'.

(Continued...)

Opinions on Mobile Web 2.0 by Ajit Jaokar & Tony Fish:

"If you're looking for the best source of information currently available on the subject of both Web 2.0 and Mobile Web 2.0, you have to go no further than this book."
Dion Hinchcliffe, Editor-in-Chief of the Web 2.0 Journal
and President of Hinchcliffe & Company

"In their latest book, the authors make fresh challenges on the paradigms in mobile data. You are not going to agree with it all, but it will challenge your own thinking"
Jeremy Flynn, Head of Commercial Partnerships, Vodafone UK

"Simply the most comprehensive and easily-accessible book on mobile Web 2.0 and its future potential available to date."
Dr Rebecca Lingwood CEng MIMechE, Director of Continuing
Professional Development, University of Oxford"

Mobile Web 2.0 **by Ajit Jaokar & Tony Fish**
futuretext 2006
ISBN : 0-9544327-6-2
Available now at major booksellers, Amazon and directly from www.futuretext.com